TELANGIECTASIA

PROCEDURES TO REMOVE DILATED
BLOOD VESSELS USING THE BLEND METHOD

by
Michael Bono

TORTOISE PRESS
Santa Barbara, California

TELANGIECTASIA:
procedures to remove dilated blood
vessels using the blend method

Published by:

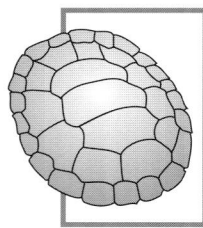

Tortoise Press
1310 San Miguel Avenue
Santa Barbara, California 93109-2043 U.S.A.

All rights reserved. No part of this book may be reproduced or transmitted in any form or by any means, electronic or mechanical, including photocopying, recording or by any information storage and retrieval system without written permission from the author, except for the inclusion of brief quotations in a review.

Copyright © 1991 by Michael Bono
Copyright Registration Number: 509-759
First Printing: 1996
Printed in the United States of America

Notice to copiers: *All of this material is copyrighted. No copying for any use whatsoever may be done without permission from the author. Permission will be given in some cases for limited photocopying, a token fee may be charged. However, you must first write to the author, at Tortoise Press, to request permission. Unauthorized photocopying may result in prosecution.*

Library of Congress Catalog Card Number: 91-92839
International Standard Book Number (ISBN): 0-9642682-1-3
Bono, Michael
**TELANGIECTASIA:
procedures to remove dilated blood
vessels using the blend method**
Bibliography:
Includes glossary.
1. Authorship and Illustrations: Bono, Michael

FOR HELEN

ACKNOWLEDGMENTS

A very special thank you to John Chapple, MD (plastic surgery), and Terry Perkins, MD (cosmetic surgery), for providing technical information for this text. Both have extensively analyzed the blend technique for removal of vascular blemishes, and have been very supportive of my efforts to present this text.

I am especially grateful to William Montagna, Ph.D. (dermatology), for reviewing this text. I appreciate his valuable suggestions and willingness to share his knowledge and materials. Dr. Montagna is the author of several noted books and scholarly articles in the field of dermatology. These titles include: *The Structure and Function of the Skin* (1974), and *Skin: Your Owner's Manual* (1985).

I would also like to thank David Davidson, MD (vascular surgery), and Charles Hamilton, MD (dermatology), for reviewing my work and making valuable suggestions and corrections. Both physicians have been eager to share their knowledge, and very supportive of my efforts in presenting this text.

Michael Bono

ABOUT THE AUTHOR

Michael Bono has been a practicing California electrologist since 1975. Formerly, he was a high school history and special education teacher. Michael works with Martin Trujillo, MD, and Terry Perkins, MD, in their Santa Barbara offices where all telangiectasia patients are medically reviewed before treatment.

Michael has taught blend classes in the U.S., Canada, Japan, Holland, Germany and Peru. He is the author of *The Blend Method: The Illustrated Manual of the Blend Method of Electrolysis* (1995). Since 1980, Michael has perfected the blend technique for the removal of telangiectases.

> *In today's world, we have become accustomed to exciting "new technology" solving our health problems. The laser especially has become the latest "wonder" of modern society. Yet the laser is a technology that is not suited for every application. For example, most surgeons have come to realize that laser is not better than the scalpel for many surgical procedures. Similarly, laser has not been shown superior for hair removal over standard electrolysis methods. In the same way, using the blend method of electrolysis for telangiectasia removal is a case of "old technology" preferable to "new technology." Simply put, you don't need a super-expensive laser to remove tiny telangiectasia vascular blemishes. An inexpensive blend apparatus not only matches laser, but in many cases is better than laser! — Mike Bono*

CONTENTS

TELANGIECTASIA: DEFINITION	15
COMMON CAUSES	17
MEDICAL CAUSES	18
POPULAR THEORIES	20
COAGULATION PROCESSES	24
SPIDER VEINS: LEGS	28
HYFRECATOR	31
LASER	32
HIGH FREQUENCY EPILATOR	33
BLEND PROCEDURES	36
BASIC PROCEDURE	38
MODIFIED BASIC PROCEDURE	41
EXTENSIVE CONDITION	43
SPIDER TELANGIECTASIA	44
DOT-TELANGIECTASIA	47
"BLOOD MOLE"	48
BLOTCHY AREA	49

DC ONLY PROCEDURE	50
REVIEW AND RECOMMENDATIONS	51
POSITIONING	52
LOCAL ANESTHETIC	53
ASPIRIN	54
OVERHEATED SKIN	55
CATAPHORESIS	56
CRUST FORMATION	56
AFTER-CARE	57
HYPERPIGMENTATION	58
LIGHTING	59
PREVENTION ADVICE	60
QUICK REFERENCE CARD	63
GLOSSARY	69
INDEX	85
BLEND METHOD	89
TEST YOUR KNOWLEDGE	93
REGISTRATION	100

FOREWORD

The procedure explained in this book is a significant improvement in the removal of telangiectases. The hyfrecator, commonly used to remove these lesions, usually produces unwanted scar tissue. Laser is popular at the moment, but the equipment and treatments are very expensive. The blend procedure is vastly superior to the hyfrecator and rivals laser! With the blend, nearly 100 percent of our cases are successful, without scarring or marks!

Michael Bono has thoroughly investigated the subject of "vascular blemishes." His description of the blend procedure is meticulous. Working for me at the Wellness Clinic in Santa Barbara, Michael has a near 100 percent success rate with my patients. The reader should be able to perform this treatment by following the instructions. Having worked with Michael for a number of years, I must express my total confidence in the blend procedure. I hope physicians and other medical professionals adopt this important methodology.

Certain nonmedical professionals have expressed concerns about potential scarring from this procedure. For example, some electrologists have stated that since this treatment effects the epidermis, scarring may result. Actually, just the opposite is true. Scar tissue does not form in the epidermis, rather scarring forms in the dermis; especially when the full-depth dermis is excessively coagulated.

Blend treatment of telangiectases is limited to the epidermis and superficial dermis, thus scarring from this procedure is nearly impossible! By contrast, the removal of hair follicles with any electric needle modality poses a far greater risk of scarring, because the full depth of the dermis is treated. Thus, treatment with the blend to eradicate telangiectases is profoundly safe: only the epidermis and superficial dermis are affected. Scar tissue never forms in the epidermis, and the coagulated superficial dermis typically heals without any visible scar or mark whatsoever.

Martin Trujillo, MD

DISCLAIMER

For the United States, this text is intended for physicians such as dermatologists or plastic surgeons, or technicians who are working directly under medical supervision, where such working arrangements are legal.

In some states and countries, only a medical doctor is allowed by law to treat telangiectases with the electric needle. In many countries (most European countries), certain nonmedical therapists, such as electrologists, are allowed to treat telangiectases. Each regional authority has specific rules governing what can and cannot be done by nonmedical therapists. Therefore, if you are not a medical doctor, you must check with state, provincial or national governing bodies and medical authorities before treating telangiectases with any electric needle device!

This text represents the experiences of several therapists and medical doctors who have removed telangiectases for many years. This information, however, is not a complete discussion of vascular diseases or conditions that cause telangiectasia. Those who wish to treat "vascular blemishes" should consult standard medical textbooks and clinical reports dealing with blood vessel diseases. Special attention should be given to texts that discuss electrocoagulation of blood vessels.

This text makes no claims or guarantees of treatment success. The suggested treatment procedures being presented are based on professional experiences. The physician must evaluate each patient and make proper medical decisions. Any misuse of the techniques discussed in this book is the responsibility of the doctor or other person performing the treatment.

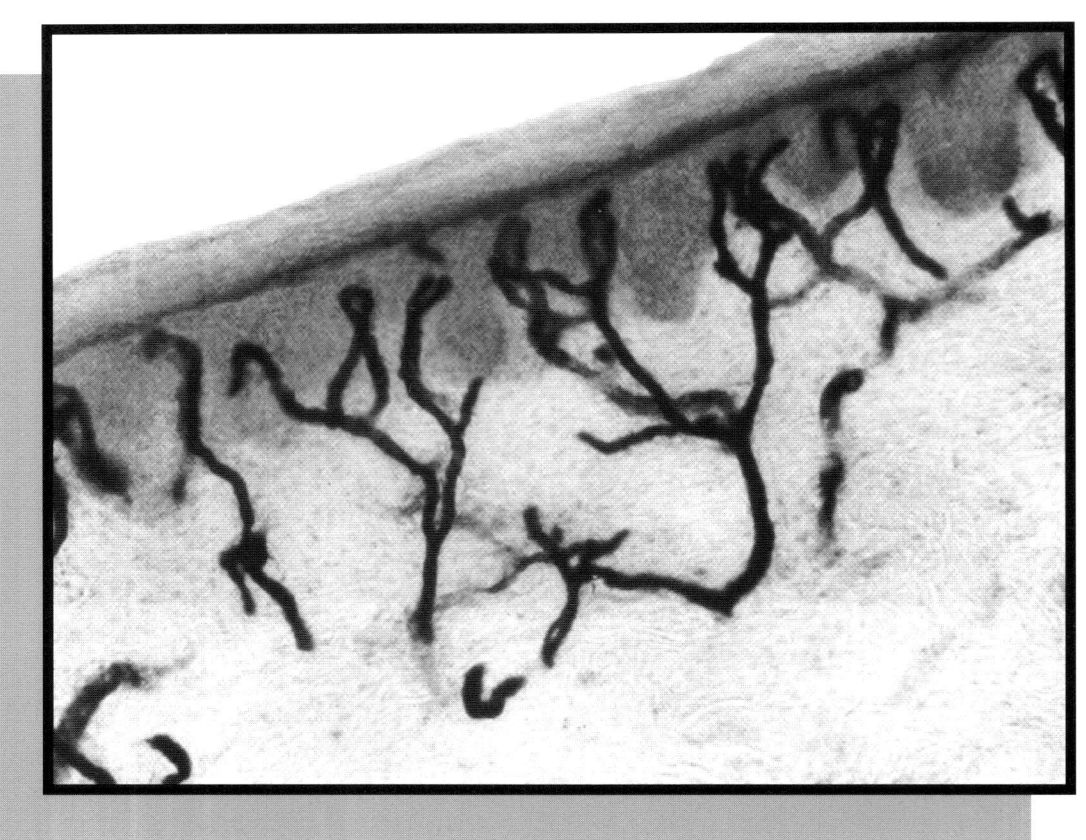

Normal capillary loops under the epidermis of the face. From this capillary bed, dilated blood vessels develop. (Photo is reprinted from *Skin: Your Owner's Manual*, by Dr. William Montagna, Ph.D., permission by William Montagna.)

TELANGIECTASIA

The blend method offers a unique and preferable treatment procedure to eliminate skin lesions commonly called "broken capillaries" and "spider veins." The medical name for these unsightly blood vessels is telangiectasia.[1] Telangiectases are small, permanently enlarged capillaries, venules, small arterioles or small veins that lie in the upper dermis and are usually dull to bright red. Normally they are linear, but they may also be dot-like markings. Some of these blood vessels are called spider telangiectasia because they form small clusters that resemble a spider or a star (Figure 1).

Except in certain disease cases, telangiectases are most often confined to the skin. Telangiectases are commonly found on the face and legs but rarely elsewhere on the body. They are often concentrated on the nose, cheeks, lips and "V" of the neck and chest. From a distance of a few meters, the person with this condition may appear to have a red nose or rosy cheeks. The neck may resemble red chicken skin.

Figure 1
Thin delicate skin under the eyes is susceptible to telangiectasia formation. The nose and cheeks are also common areas for these lesions. Illustration shows common spider telangiectases.

---

1) Usage of the term **TELANGIECTASIA** (tel ANGE ek TA sia):
a) *AS A CONDITION:* The patient suffers from telangiectasia. I am studying telangiectasia. (The term "telangiectasis" is also used to describe the condition.)
b) *SINGULAR:* The patient has one large telangiectasia.
c) *PLURAL:* I have three telangiectases. ("telangiectasias" is also used for the plural, although less frequently.)
d) *ADJECTIVE:* The injury caused a telangiectatic condition.

The blend method of electrolysis unites DC current and HF current in a single needle—the currents are applied together simultaneously. For telangiectasia removal, 0.2 to 0.4 milliamperes of DC, and 40 to 55 volts of HF are used. When very low levels of DC and HF are used together in this way, they are said to be used as "face-technic." Because all blend devices have different numbers for the HF settings, *you must refer to your manual to correctly set the HF and DC levels of your blend device.*

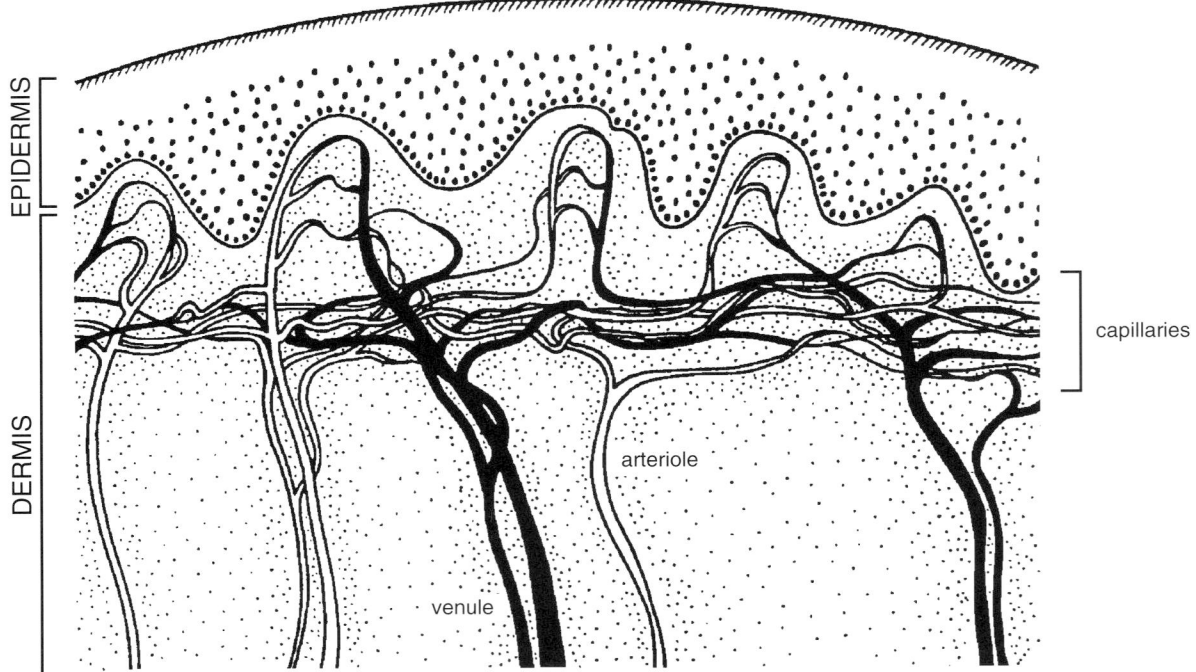

Figure 2
The upper (papillary) dermis contains an intricate network of blood vessels; these are arterioles, capillaries and venules. Capillaries loop upward and nourish the epidermis. Telangiectases form when vessels of the upper dermis become permanently dilated. The above illustration depicts normal blood vessels of the skin.

> The Greek word *telangiectasia* is made up of root Greek words that are common to the English language. For example, *telos* (meaning at the end or distance) is used in such words as telogen (at the end of growth), television, telescope and telephone. The word *ageion* (pertaining to blood), is used in the words angiogenesis (growth of blood vessels), and angiology (the study of blood—circulatory—system).

Close examination of these ruddy areas reveals dilated blood vessels placed randomly. It is this fragmented appearance that makes the blood vessels look broken. The vessels, however, are not actually broken. A broken vessel would bleed under the skin and form a bruise, which is not the case with telangiectasia. The term "broken capillary" is therefore incorrect.

The medical term itself describes the nature of these unattractive blood vessels. The Greek word *telangiectasia* means "end vessel dilation." *Telos* means "end," *angeion* pertains to a blood vessel (or receptacle) and *ektasis* means "stretching out." Accurate descriptions of telangiectases are: thread veins, blood spots, dilated veins, spider naevi, spider veins, thread capillaries, dilated capillaries and vascular blemishes. Whatever physical form the telangiectasia takes, it is by definition a dilated blood vessel at the end of the circulatory system where cells are nourished.

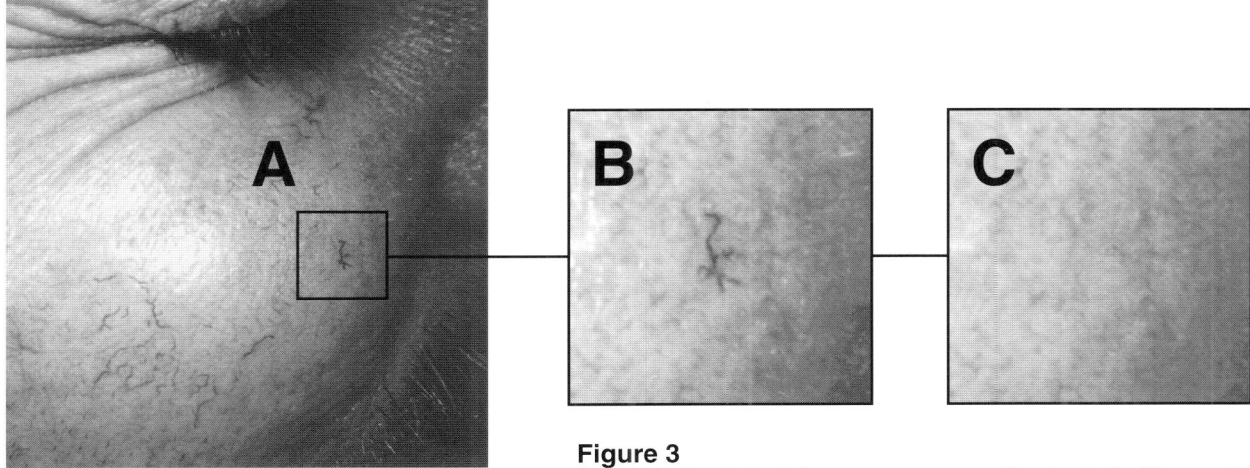

Figure 3
Common telangiectasia from sun exposure on the cheek of a 42-year-old Caucasian man. A) Cheek before blend treatment. Note telangiectasia in square. B) Photographic enlargement of telangiectasia before treatment with the blend. C) Same area 8 weeks after one treatment with the blend.

There are many different causes of telangiectasia. A telangiectatic condition can be minor or severe with no known cause or may be induced by a specific illness. But whatever the cause, telangiectasia is only a condition; it is not a disease.

COMMON CAUSES

The most common non-disease related causes are: aging, X-ray,[2] sun and light exposure, normal wear and tear on the skin and injury, such as an abrasion (Figure 3). Such common injuries as improper eyebrow plucking can pinch the skin and dilate capillaries. In this case the capillary is squeezed beyond its elasticity, almost like overinflating a balloon. Carelessly squeezing out a comedo (blackhead) is a similar injury that can cause telangiectases. Injury-induced telangiectases usually respond well to treatment (Figure 4, next page).

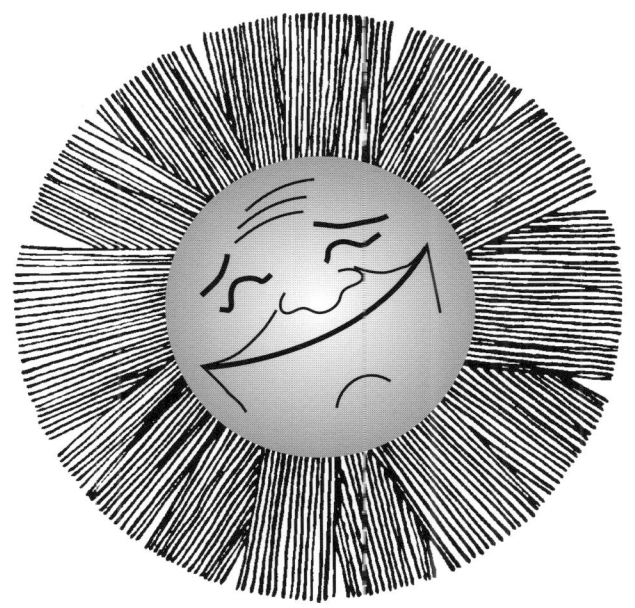

Excess sun exposure is unquestionably the foremost environmental factor causing telangiectases in predisposed individuals. Nearly all fair-skinned people are prone to telangiectases. These lesions gradually show up with age, usually in the mid-40s.

---

2) X-ray was once used to treat acne and other skin conditions. Some who received this treatment developed telangiectasia. There does not seem to be significant risk of getting telangiectases from dental, chest or other diagnostic uses of X-ray.

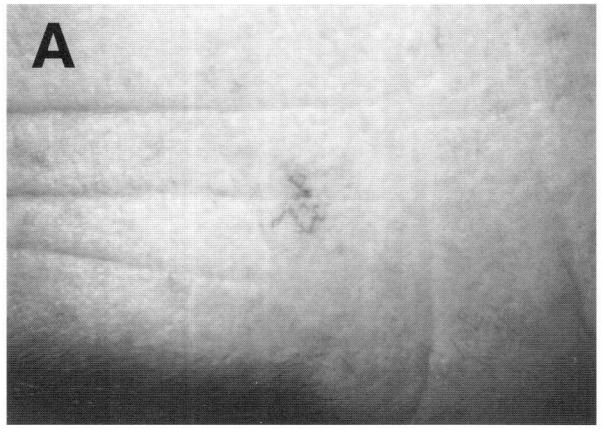 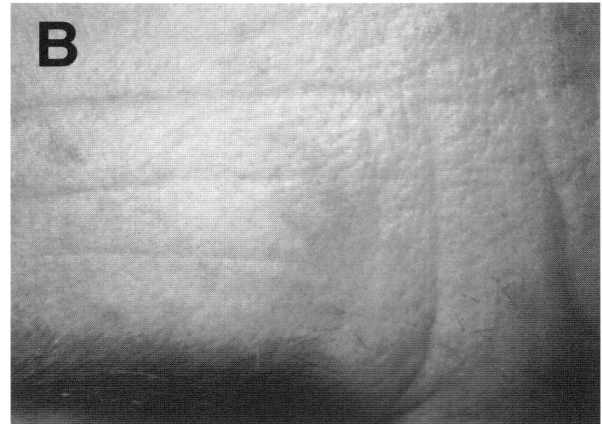

Figure 4
The woman above said that when she opened her refrigerator, a 10 pound (4.5 kilo) frozen cheese fell out on her forehead and gave her a bruise. Afterward, she claimed, a telangiectasia formed. A) Telangiectasia before treatment. B) Vessels after 2 treatments. Injury induced telangiectases are easily removed and do not return.

TELANGIECTASIA IN WOMEN

Telangiectasia occurs more frequently in women than in men. Spider veins on the legs are common to women but rarely found on men. Telangiectases sometimes appear during pregnancy but usually disappear six weeks after delivery. However, the same areas of blood vessel enlargement often reappear during a subsequent pregnancy.

> It's a good idea to not treat a pregnant woman. Should there be any complication or anomaly with the baby, suspicion might be cast on the electrical currents of the treatment. You could face a lawsuit. My suggestion is to not treat pregnant women!

Increased hormone production during pregnancy causes increased blood pressure and volume which may then produce temporarily dilated capillaries. Do not treat a patient's telangiectases during or soon after pregnancy because the problem usually disappears naturally.

MEDICATION AND SURGERY

Medication can cause dilated blood vessels of the skin. Oral contraceptives containing estrogen (female hormone) can elevate normal hormone levels and cause telangiectases. Long term use of Retin-A for cosmetic purposes can dilate facial capillaries in certain individuals.[3] Chemical peels of the face to diminish wrinkles and age spots can also bring about the condition.

---

3) Retin-A (generic name: tretinoin) is an externally applied medication for acne, but is also used to improve the appearance of sun-damaged skin. The drug causes increased blood flow to the skin. Some patients may develop dilated capillaries from continuous use which may or may not be permanent. (Refer to the current issue of *Physicians' Desk Reference*.)

Telangiectases sometime form in scar tissue. Surgical removal of skin lesions such as basal cell carcinomas or large moles from the face can produce scar tissue with visible telangiectases. Patients who have had face lift surgery occasionally develop telangiectases along the ear where an incision was made (Figure 5). Telangiectases in scar tissue can be removed, which then reduces the noticeability of the scar itself. If keloids are seen prior to treatment, consult with the doctor because treatment could worsen the condition.[4]

HEALTH PROBLEMS AND DISEASE

Patients with high blood pressure often have dilated facial capillaries; their skin can appear quite red. A physician must decide if treatments will benefit such patients. The condition can be temporarily improved but not cured. Dilated vessels will continue to form because of the high blood pressure.

Telangiectases may indicate health problems and can therefore be a crucial diagnostic indicator. Disease may be the cause if the case is extensive or recent in development. Standard textbooks of dermatology list twenty-three diseases that can cause telangiectases.[5]

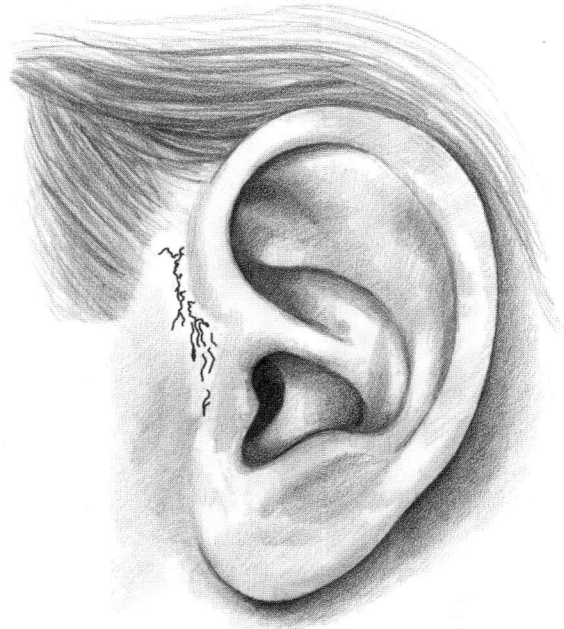

Figure 5
Telangiectases can form along the ear following cosmetic face lift surgery.

---

4) Keloids are elevated fibrous tissue growths associated with scars. The tendency to form keloids is hereditary.
5) Rook: *Standard Textbook of Dermatology*. (This text is updated periodically, refer to current issue.)
Thomas Goodman: *The Skin Doctor's Skin Doctoring Book*.

Such diseases as scleroderma, dermatomyositis, lupus erythematosus, chronic hepatitis and Raynaud's phenomenon can manifest telangiectasia. Telangiectasia may be seen on late-stage AIDS patients.[6] Telangiectases can be associated with skin diseases and chronic skin conditions such as rosacea (a persistent inflammatory facial eruption).

Most patients complaining of telangiectases have minor cases that are not disease related and are of cosmetic importance only. However, absolutely every patient must be medically evaluated before treatment to rule out disease! If a disease or medical problem is present, a physician must decide if treatment will be successful.

POPULAR THEORIES

There are several popular theories regarding the genesis of telangiectases that are medically unproved, yet might be tenable. One theory places the blame on rapid changes in environmental temperature and another places the blame on alcoholism. Because these two theories are widely believed, a short discussion follows.

---

6) Chyang T. Fang, Ph.D.: "HIV Testing and patient counseling," *Patient Care*, Oct. 30, 1989, page 24. (Telangiectases may be seen on the ear, hand, foot or trunk of late-stage AIDS patients.)

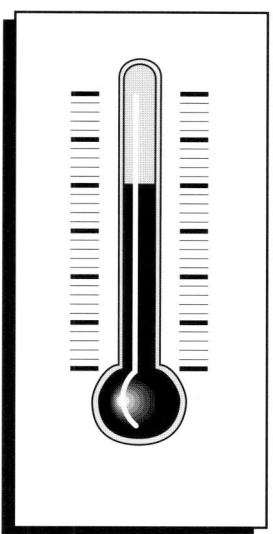

THE TEMPERATURE THEORY

A primary function of the blood in the skin is to regulate body temperature. In a hot environment, blood rushes into the skin; sweat forms and evaporates. This evaporation cools the skin and blood, and lowers body temperature. Because of the increased blood volume, our skin appears florid.

By contrast, in a cold environment the blood moves out of the skin and into the internal organs to protect them from the cold. This is why our skin turns pale in the cold outdoors of the winter season.

The "temperature theory" holds that when rapidly going from a cold to a hot temperature, the blood rushes so quickly into the face that capillaries burst and permanently dilate. Supposedly, going from the cold outdoors of winter into a heated building or sauna causes this phenomenon. This theory points to people living in cold climates, such as found in northern Europe, where temperature extremes are common. In fact, telangiectases are seen more frequently on northern Europeans than southern Europeans.

Certainly, capillaries of the skin enlarge in warm temperatures to allow increased blood flow. However, there is no clinical evidence that they actually burst under any natural temperature change condition. It seems doubtful that capillaries of the skin remain permanently dilated from rapid temperature changes. Healthy capillaries are amazingly elastic; they expand and contract continually.

While the temperature theory seems plausible, especially since the condition affects northern Europeans frequently, there might be a simpler explanation. If a person has little pigment in his or her skin, blood vessels are more visible through this transparent skin. If the person also has thin skin, blood vessels will be more prominent, especially on the nose and cheeks where facial skin is most thin. Thus, telangiectases, whatever the underlying cause, appear more frequently on white, thin-skinned people. This is why the condition is often seen in northern Europe where many inhabitants have thin white skin.

All blood vessels of the skin do not react consistently to temperature stimuli. Most blood vessels in the skin contract under cold conditions, however, some vessels become dilated and flushed with blood. These areas, called the "blush areas," give us our rosy cheeks and noses in winter.

Telangiectases often form in the blush areas. A person exposed to extended periods of cold may develop telangiectases in the blush areas. Skin and blood vessels lose elasticity with age and eventually these dilated blood vessels could become permanent. Men who work outdoors in cold climates often have facial telangiectases.

> Under certain types of psychological stimulation, such as embarrassment, the blush areas become dilated. These areas include the cheeks, head, neck and upper chest. In his book *Skin: Your Owner's Manual*, Dr. Montagna states that this psychologically caused dilation is "peculiar because it is associated with an adrenaline response." He points out that dilation of blood vessels is normally controlled by acetylcholine (antagonist of adrenaline).

ALCOHOLISM THEORY

Another popular theory about telangiectases argues that excessive drinking of alcohol causes "broken capillaries." The red-nosed drunkard is the standard example. This theory appears supportable because alcohol consumption does cause blood vessels to dilate. As capillaries expand from alcohol, the blood supply increases which makes the skin appear redder and feel warmer to the touch.

Because people often witness this temporary reddening of the skin caused by alcohol, they tend to associate enlarged facial capillaries with alcoholism. Although alcohol might cause permanent capillary enlargement in certain individuals, it seems the effect of alcohol on the skin is mostly temporary. Medical literature does not include moderate drinking as a prominent cause of telangiectasia formation.

Many years of incessant drinking, however, might cause telangiectases. Since blood vessels and skin lose elasticity with age, the vessels could eventually remain dilated. Prolonged and excessive drinking can also damage the liver. This damage could interfere with the liver's ability to control hormone levels and cause telangiectases. It is notable, however, that dark-skinned alcoholics usually do not have telangiectases whereas white-skinned people who do not drink at all sometimes have the so-called "drunkard's nose."

HEREDITY

Heredity seems to play the most significant role in the development of telangiectasia in the completely healthy person. Thus, the condition can be a family trait. Whatever the underlying cause, the condition most definitely affects white, "thin-skinned" people more frequently than dark, "thick-skinned" people.

Inherited telangiectasia occurs in up to 15 percent of the completely normal healthy populace. Nearly all medical texts define hereditary telangiectasia as "a congenital weakness in the blood vessel walls."[7]

When treating patients with telangiectases, do not simply assume that they are alcoholic or that they have been careless with their skin. Explain to them that heredity is the primary factor that causes a telangiectatic condition in the completely healthy person.

Patients often blame themselves for their condition and are embarrassed because society equates these lesions with alcoholism. Because of this feeling, they often have contrite explanations as to why these blood vessels developed. By explaining that the condition is hereditary, their embarrassment can be relieved and self-respect improved.

Each culture holds different beliefs as to the cause of telangiectasia. Keeping in mind that heredity plays a significant role in the development of these lesions, how many of the following ideas do you think are correct?

In Holland, famous for Indonesian food, many believe that spicy food can cause dilated capillaries. (Some of those exotic dishes certainly can cause a rush of blood to the face!) The Dutch sometimes use the term "butcher's cheeks" to describe telangiectases on the cheeks. They believe that working in a cold freezer causes the condition. In Germany many believe that drinking too much Schnapps (an alcoholic beverage) causes the condition. A bright red nose is called a "Schnappsnase."

During classes in Japan, several students told me they had seen the condition on fishermen living on the cold northern island of Hokkaido. The students thought the cold weather was responsible. During classes in Peru, I found that few students had ever seen telangiectases. The condition is rare on people of native heritage. On the last day of class, however, a woman came for treatment who had an extensive case. Although born in Peru, she was of European heritage and had very white skin.

---

7) Moschella and Hurley: *Dermatology* (W.B. Saunders, 1985).

PROCEDURES TO REMOVE TELANGIECTASES

COAGULATE THE VESSEL

All treatments to remove telangiectases attempt to block the flow of blood in the vessel. Blockage of the capillary is achieved by coagulating a small segment of the vessel. The blood is "clotted" at the treatment point. The clot blocks the flow of blood.

There are two ways the capillary heals and becomes functional again following treatment; either way the vessel becomes less visible. First, the damaged vessel and clot are permanently replaced by scar tissue; blood then flows through adjoining vessels. Second, a small channel, or canal, forms through the scar tissue. In this case the vessel is again functional, but smaller and less visible. This process is called recanalization (Figures 6 to 9).

Dilated capillaries of the upper, or superficial, dermis are essentially dysfunctional. They carry blood, but are too large to nourish skin cells effectively. Therefore, proper removal of telangiectases causes no risk to the health of the skin. When telangiectases are removed, new normal-size capillaries regrow to efficiently feed the skin. This process is called angiogenesis.

> Can the coagulated segment (clot) dislodge and form an embolism in the blood system, perhaps settling in the heart or lungs? This seems unlikely since surrounding tissue is coagulated along with the capillary itself. In this way, the clot is held in place by surrounding coagulated tissue. Also, arterioles and capillaries become smaller in diameter as they reach the cells. Any loose coagulated material would therefore flow into a structure of diminishing diameter and be held until absorbed by the body. The risk of embolism is negligible when coagulating minute telangiectases.

Figure 6
Dilated capillary. Arrows show direction of blood flow.

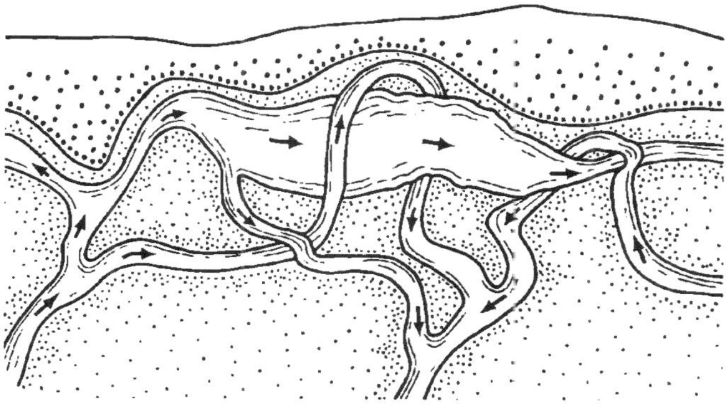

Figure 7
Blood vessel is coagulated.

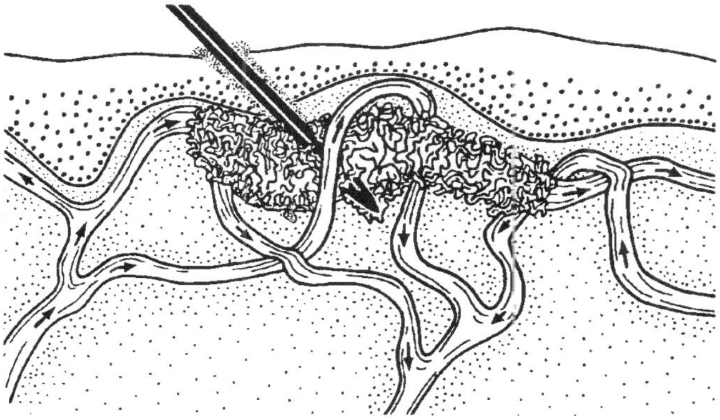

Figure 8
In this case the blood vessel is completely replaced by scar tissue. Blood flows through connecting vessels.

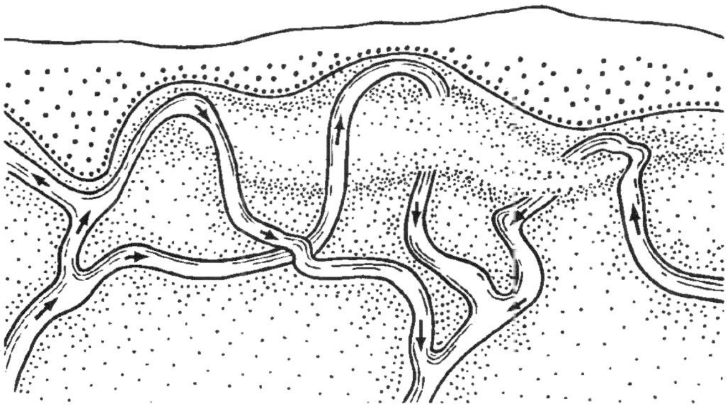

Figure 9
In this case blood forms a channel (or "canal") through the scar tissue forming a new, but smaller, vessel. This process is called **recanalization**.

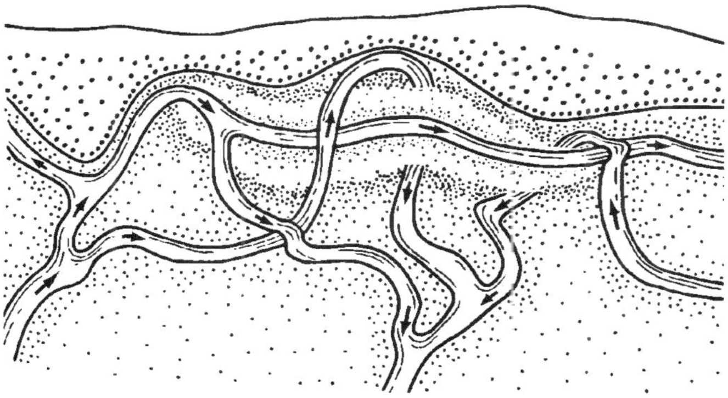

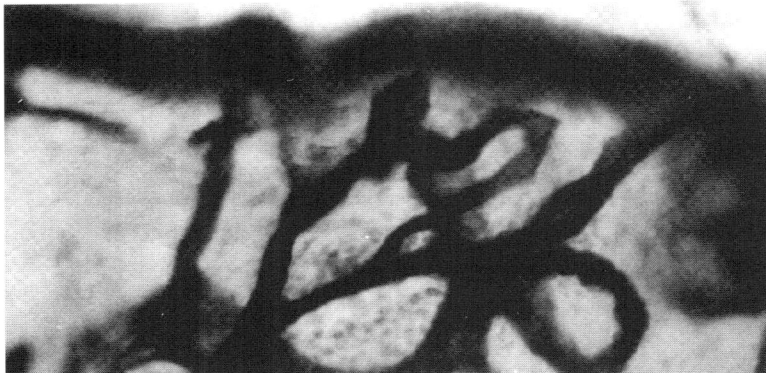

Figure 10
Intricate capillary loops in the lip. Sometimes loops become dilated and look like tiny blood droplets. Such telangiectases can be removed. (Photo courtesy of William Montagna, Ph.D.)

VEINS AND CAPILLARIES

Think of the coagulation treatment as constructing a barrier against the flow of blood in the vessel, much like a dam holds back a river. A river downstream of a dam dries up, as does the blood vessel. This process works well against capillaries but not against veins or venules because of the difference in the direction of blood flow in the respective vessels.[8]

Capillaries in the skin branch out from arterioles and become progressively thinner as they carry blood to the cells. A single clot formed at point A in the illustration eliminates the connected "downstream" capillaries (Figure 12). Venules, however, join into progressively larger veins as they return blood to the heart and lungs. A single clot formed at point B in the illustration would not cause removal of connecting venules and veins (Figure 13). Therefore, for the treatment to be effective, every venule and vein in the area must be individually coagulated. Practical experience shows that numerous clot formations are required to remove even a single vein or venule. The process is lengthy.

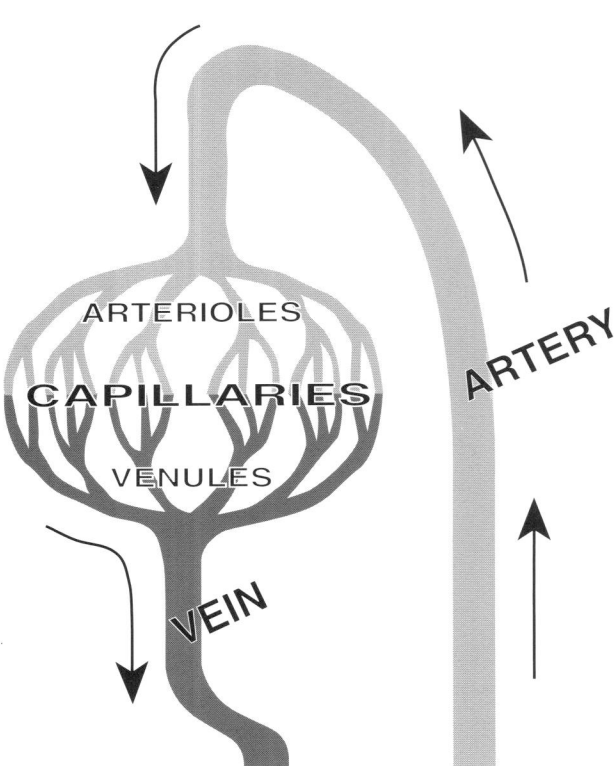

Figure 11
Arteries carry nutrient and oxygen enriched blood from the lungs and heart to the cells. Arteries branch out to form small arterioles and finally tiny capillaries. At the capillary level, cells are nourished. Oxygen and nutrient deficient blood are then returned to the heart and lungs by venules that branch-out into ever larger veins. Veins return blood to the heart and lungs.

---

8) The venous vessels normally treated by electrocoagulation are actually venules (small veins). Medical literature, however, often simply refers to treatable venules as "veins."

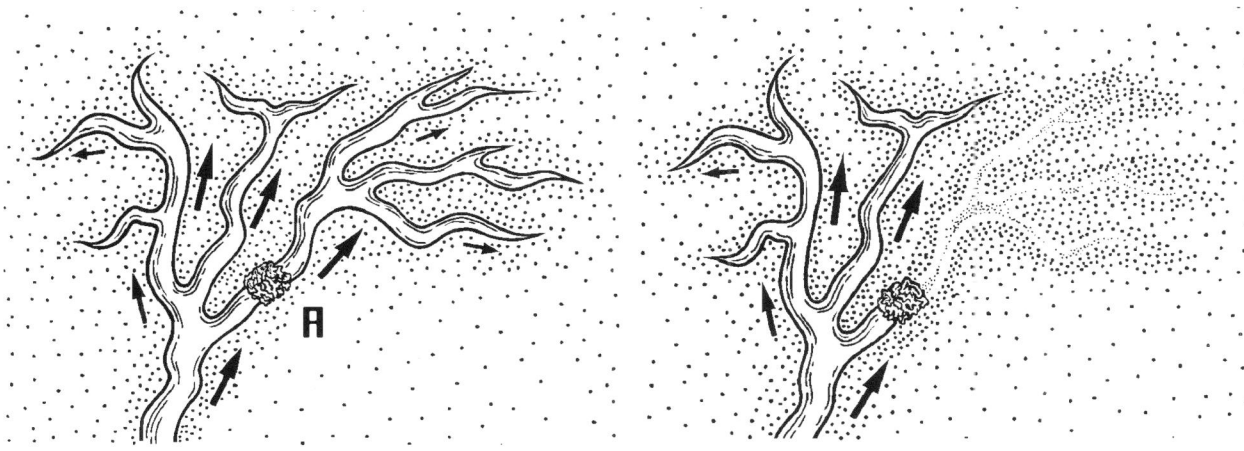

Figure 12
If the blood vessel is a **capillary**, a coagulated clot (point A) eliminates the connected "downstream" section of the vessel. Arrows show direction of blood flow.

Figure 13
If the blood vessel is a **venule**, a coagulated clot (point B) eliminates only a small segment of the "downstream" vessel. Arrows show direction of blood flow.

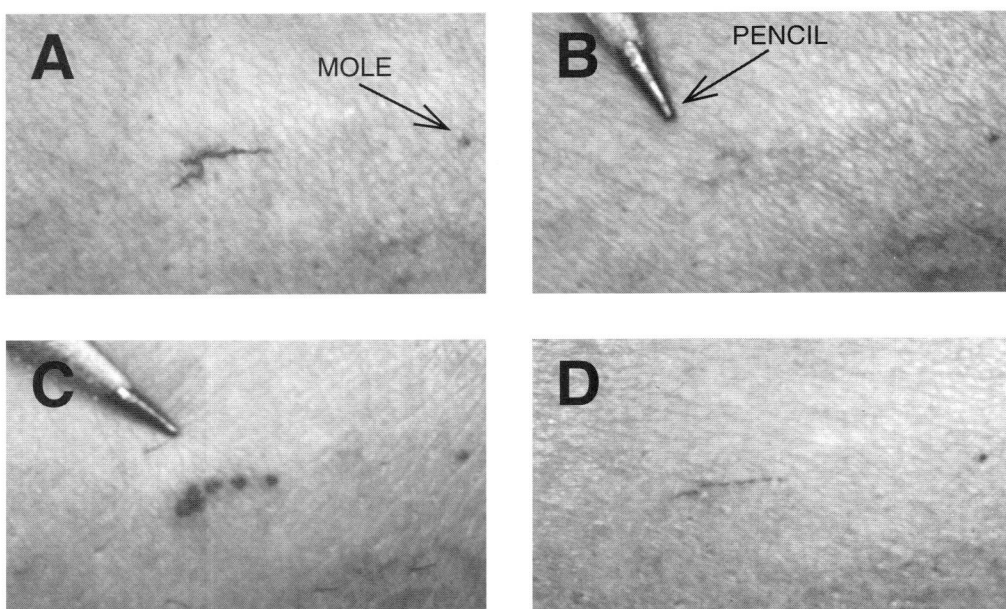

Figure 14
Spider vein on the upper leg of a 50-year-old man. A) Before treatment. B) Immediately after treatment. Vessel appears completely coagulated. C) Large crusts after about 10 days show that significant current strength was used to coagulate vessel. D) About three weeks after treatment. Notice that the vessel appears somewhat smaller. One or two more treatments will be necessary to remove the vessel. However, even after several treatments, leg telangiectases may remain. Blend treatment of even minute leg vessels is seldom successful.

SPIDER VEINS: LEGS

Telangiectases on the legs are often venules or veins and are therefore very difficult to remove by any electrocoagulation method. Blood pressure in the legs is greater than the face and increases as the patient stands up and walks. Therefore, it is most difficult to keep the coagulated segment in place following treatment. For these considerations, treatment of leg telangiectases by electrocoagulation usually fails (Figure 14).

By contrast, facial telangiectases are much easier to remove by electrocoagulation because they are thinner, usually capillaries and have lower blood pressure.[9] For this reason, treatment of telangiectases with the blend method should primarily be limited to the face, neck and upper body (back, chest and abdomen).

Refer your patients with unwanted leg vessels to a physician who specializes in the removal of spider veins and varicose veins of the legs. Standard treatments include laser, surgery and sclerotherapy.

> Treating veins on the legs, especially vessels associated with varicose veins, could be dangerous. The removal of these vessels could worsen an existing damaged blood supply problem. Do not treat large vessels anywhere on the body! Large vessels must be treated by medical doctors who specialize in removing these veins. Limit yourself to treating small telangiectases of the upper body and the face.

---

[9] Medical literature usually assumes dilated vessels of the upper body are capillaries.

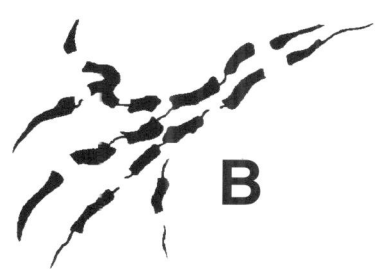

Figure 15
As tiny leg venules are removed in stages, they temporarily resemble a peculiar dotted-line.

After one or two treatments.

After many treatments.

Although treating small blood vessels on the legs is usually futile, sometimes it does work. If you attempt leg work, advise the patient that treatment success is typically very poor.

If you decide to treat the legs, treat only thread-size vessels on the upper legs. Vessels on the lower legs and ankles are nearly impossible to remove by any electrocoagulation method (Figure 16). Try experimenting with a few vessels to see if they can be successfully treated. Medium to high "face-technic" current levels are required for leg work.[10] Significant crusting is common (see crust size, Figure 14).

Since leg vessels are usually venules, many treatments will be necessary to achieve success. As the tiny venules are successfully removed, they often look like a dotted-line until completely eliminated (Figure 15).

Using a compression bandage (such as the elastic "Ace" bandage) immediately after blend treatment of leg vessels may somewhat improve treatment outcome. Experimentation is continuing using this technique. Preliminary results appear very promising. Essentially, the compression bandage restricts blood flow to the treated vessels during the initial healing period.

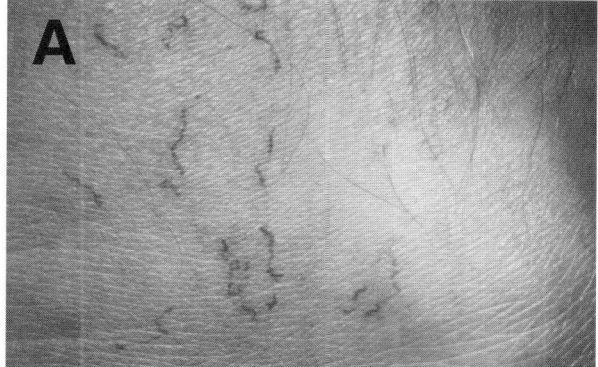
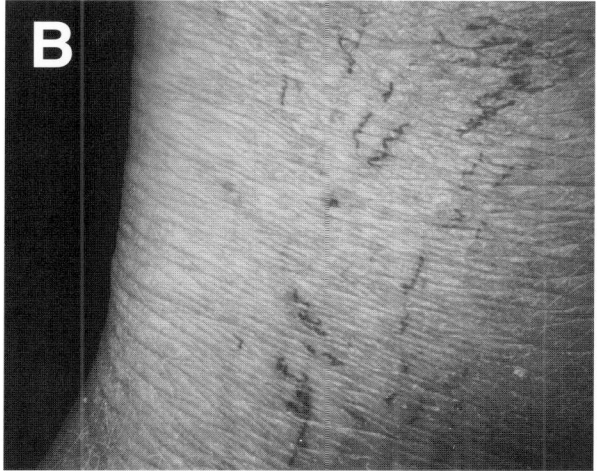

Figure 16
Telangiectases on the ankles are nearly impossible to remove with any electrocoagulation technique. Do not attempt to remove such vessels with the blend. A) Telangiectases on ankle of 42-year-old man. B) Telangiectases on ankle of 72-year-old woman.

---

10) The term "face-technic" is specific to the field of electrology. The term refers to both HF and DC currents being used simultaneously at low levels. "Medium to high" face-technic is a higher HF and DC setting than normally used for telangiectasia treatment. See *The Blend Method*, by Michael Bono, pages 131-136, and 155-164.

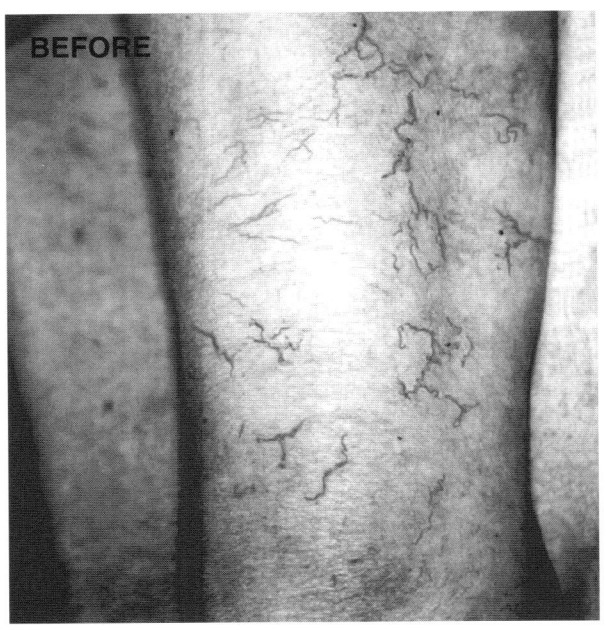 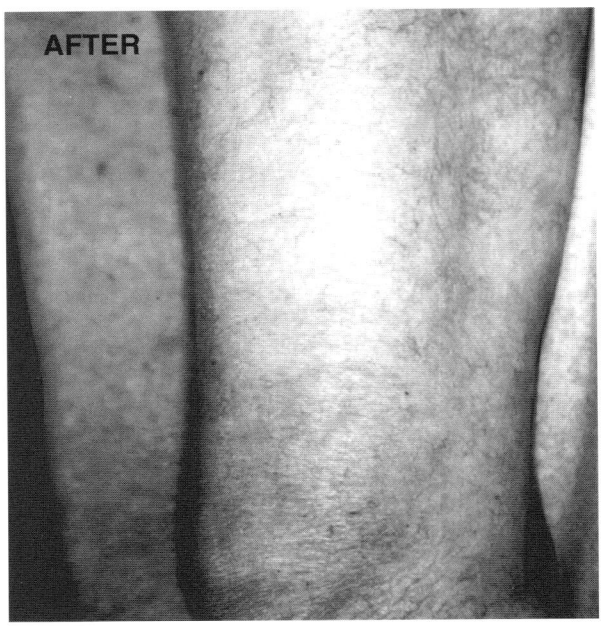

Figure 17
Removal of spider veins from the legs of a 56-year-old woman by injection of sclerosant solution. Vessels of the above size and larger, including varicose veins, are best treated by physicians specializing in such treatment. (Photo courtesy of Jay Applebaum, MD, dermatology and sclerotherapy, Orange, California.)

INJECTION-COMPRESSION

The injection of sclerosants (solutions causing sclerosis) with such trade names as Sotradecol® are used by physicians to remove spider and varicose veins on the legs. The caustic solution causes irritation of the vessel lining so that a thrombosis, and later sclerosis, develops. The leg is securely bandaged for approximately one week. The patient walks three miles per day for about four weeks. During this time, vessel walls shrink together and the vein is replaced by scar tissue. The vein is rendered dysfunctional and cosmetically invisible.

Some bruising is common with this procedure and marking can result if the physician is not skilled. For example, if the physician misses the vein and injects the surrounding tissue, unwanted skin destruction and scarring can take place. If the physician is skilled, however, removal of even large veins is usually successful with permanent results seen within a few weeks after treatment. Patients with large unwanted blood vessels should be referred to a dermatologist or vascular surgeon who specializes in this treatment (Figure 17).

Fine facial telangiectases (about the thickness of a thread) cannot be injected because such vessels are thinner than the hypodermic needle itself. Capillary walls are also very thin. Caustic solution can easily dissolve and pass through the thin vessel wall into surrounding tissue and cause unwanted skin damage. Fine upper body and facial telangiectases are best treated by the electric needle blend method.

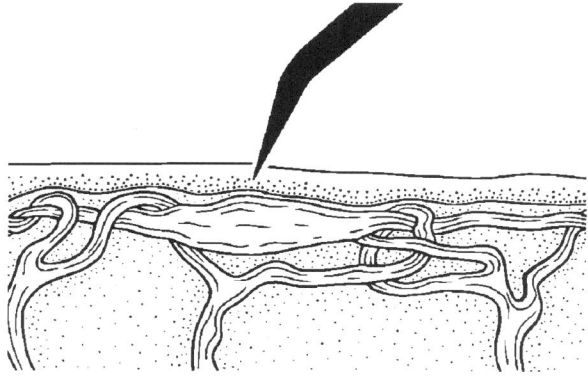

Figure 18
With the hyfrecator, the stylus is fractionally inserted into the epidermis only; not the blood vessel.

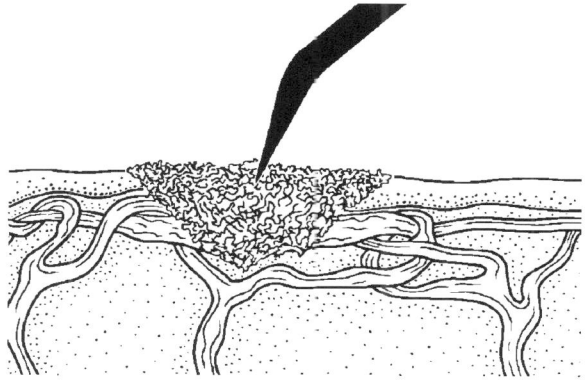

Figure 19
Too much surrounding tissue is coagulated; the dermis is significantly affected.

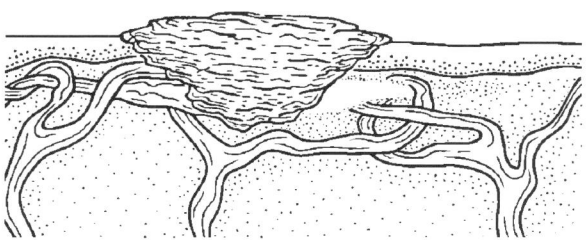

Figure 20
A large crust forms over the treated area.

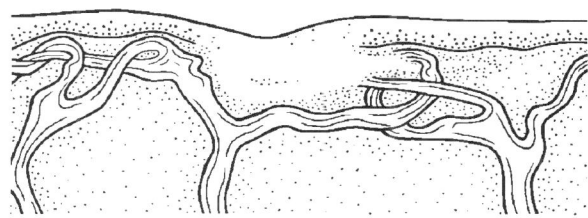

Figure 21
In most cases visible scar tissue is seen after treatment with the hyfrecator.

HYFRECATOR

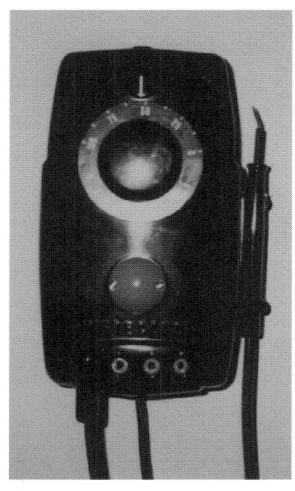

HYFRECATOR

Medical doctors use a high frequency (HF) cauterizing device called a hyfrecator to eradicate telangiectases of the face and body. (The term "hyfrecator" is an acronym of *high frequency desiccator*.) This device is used to burn off skin abnormalities such as warts and moles. The HF generating hyfrecator is a standard apparatus used by nearly all dermatologists worldwide.

The HF energy from the hyfrecator is substantially higher than the HF levels the blend machine produces.[11] When removing telangiectases, the doctor just touches the stylus point to the skin; needle insertion is not necessary to coagulate the blood vessels (Figures 18 to 21).

---

11) Certain hyfrecators produce too much HF, even at "0" on the dial. Doctors, used to burning off warts and moles, typically use too much HF current to remove telangiectases. The blend uses a tiny fraction of the current commonly used with the hyfrecator.

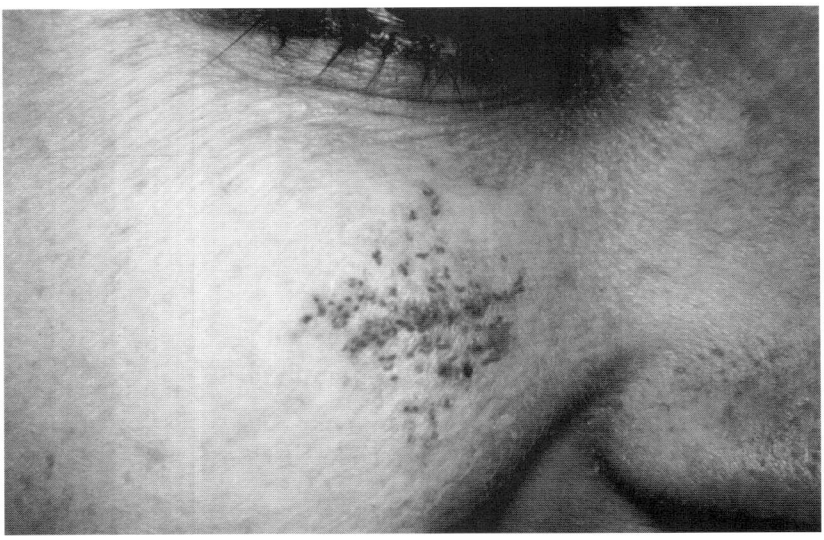

Figure 22
Normal scabbing after treatment with the hyfrecator is often severe. By contrast, post-treatment scabbing with the blend is barely noticeable (crusts are usually the size of a pin point).

With the hyfrecator, too much surrounding skin and deeper layers of the dermis are affected. Scabs of 4 millimeters or more form, and tiny scars are common (Figure 22). Sadly, the hyfrecator is the primary instrument used by medical doctors to remove these lesions. Because of the substantial overtreatment potential, most doctors are reluctant to treat telangiectases.

LASER

Laser is a high-intensity ray of light with many medical applications. There are numerous types of lasers in use today, including the argon and yellow dye laser. These laser beams are specifically absorbed by red objects and are therefore used to coagulate blood vessels. The blood absorbs the light and enough heat is produced to coagulate the blood, provided the blood flow is not too fast to dissipate the heat.

The laser beam is not absorbed by clear or white tissue and can therefore pass through the lens and fluid of the eye without damaging the lens. Thus it is used to cauterize small bleeding blood vessels of the retina at the back of the eye (Figure 23). The laser may also be used to cauterize blood vessels of the skin, as in telangiectasia, or areas of blood vessel birthmarks.

Figure 23
Argon laser is used for eye surgery. In the above photo, laser is microscopically focused for repair of the retina. (Photo courtesy Glynne Couvillion, MD, Santa Barbara.)

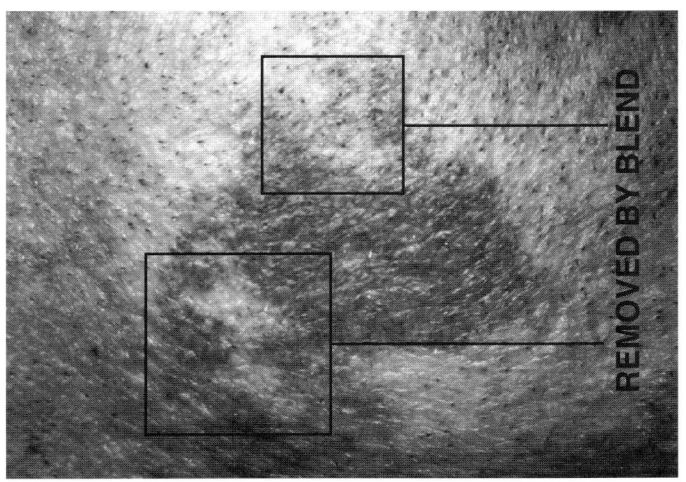

Figure 24
Port wine birthmark: vascular port wine birthmark on cheek of 56-year-old man. Although the blend can remove such marks, too many treatments are required. Port wine marks are best removed by laser. Framed areas show portions of a port wine birthmark removed by the blend.

Laser is successful in removing large vascular marks such as the "port wine" birthmark but is generally considered excessive for treating small thread-like facial telangiectases (Figure 24). Precautions must be taken so that the light beam does not enter the eye of the technician or patient. The equipment is very expensive and the cost of treatment is high. The treatment is not particularly painful, but swelling is common because too much tissue is often affected. Scarring sometimes occurs.

HF EPILATOR

An HF epilator,[12] either automatically or manually controlled, can adequately coagulate unwanted capillaries. The operator inserts a fine sharp needle into the capillary, and then cauterizes the vessel with the HF current. This method is usually better than the hyfrecator because the operator is able to select individual blood vessels for treatment.

Figure 25
A "world famous" port wine birthmark seen on Mikhail Gorbachev.

---

12) High Frequency (HF) is also known as short wave, thermolysis, radio frequency (RF) and diathermy.

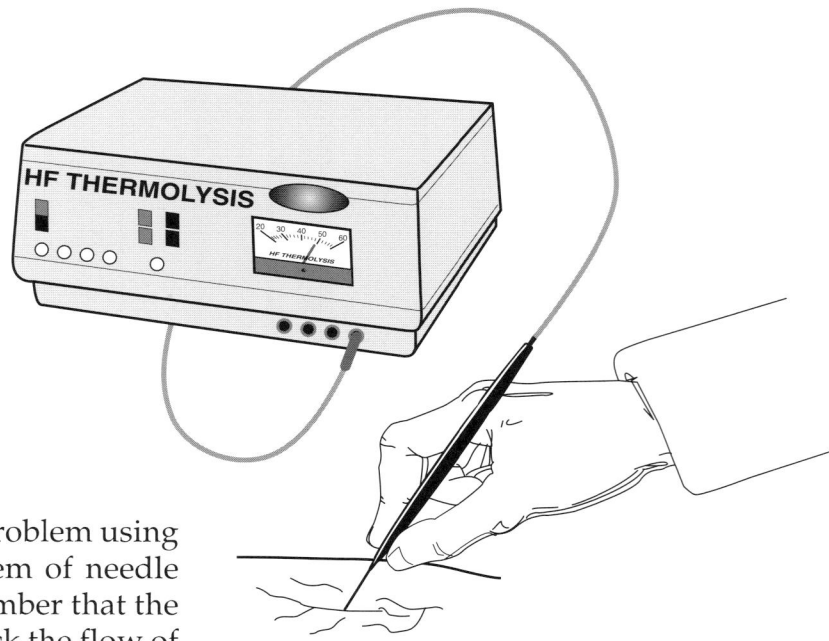

Figure 26
Tapping Technique. Operator just inserts needle fractionally into the epidermis and uses only high frequency to achieve vessel coagulation. There are unfortunate disadvantages to using this technique.

There is, however, a problem using HF current alone: the problem of needle removal from the clot. Remember that the operator is attempting to block the flow of blood in the capillary by coagulating a small segment and forming a clot. The coagulated clot is like a dam holding back the flow of blood. If the clot is inadvertently removed, the vessel refills with blood and the treatment fails.

The HF effectively forms a coagulated clot, but it also evaporates off the moisture at the needle contact point. The current is so drying that the needle often sticks to the coagulated segment. As the operator withdraws the needle, the needle pulls out the clot and the vessel reopens (Figures 27 and 28, facing page).

To correct this sticking problem, some HF operators advise simply touching the needle point to the skin surface or inserting only a "fraction" into the epidermis. In this way, the needle never enters the capillary itself and needle sticking takes place only on the epidermis. This technique is called "tapping along the dilated capillary" (Figure 26).[13]

The disadvantages of the "tapping technique" are obvious. In order to successfully coagulate the capillary through the skin, relatively high HF levels are used. Because of the high current, the dermis is unnecessarily coagulated which can result in small scars. Also, deeper vessels cannot be reached by the HF current because the needle is inserted into the epidermis only, not actually into the vessel.

The problem of the needle sticking to the coagulated clot is the main drawback of using HF current epilators. As we shall see, the problem of needle-sticking is completely eliminated by using the blend method's DC current to withdraw the needle from the coagulated segment.

---

13) Ann Gallant: *Principles and Techniques for the Electrologist*, 1983, pages 193-194.

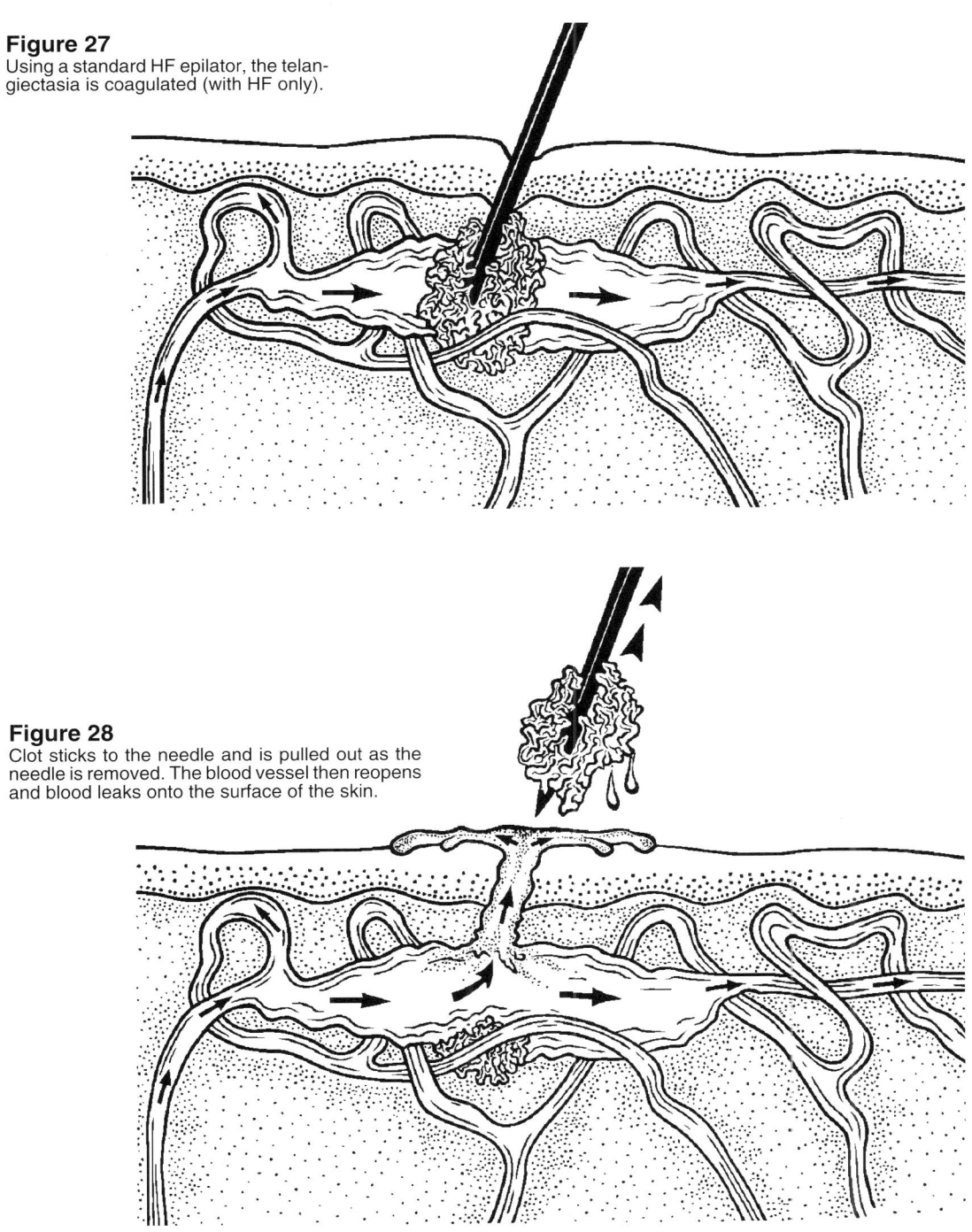

Figure 27
Using a standard HF epilator, the telangiectasia is coagulated (with HF only).

Figure 28
Clot sticks to the needle and is pulled out as the needle is removed. The blood vessel then reopens and blood leaks onto the surface of the skin.

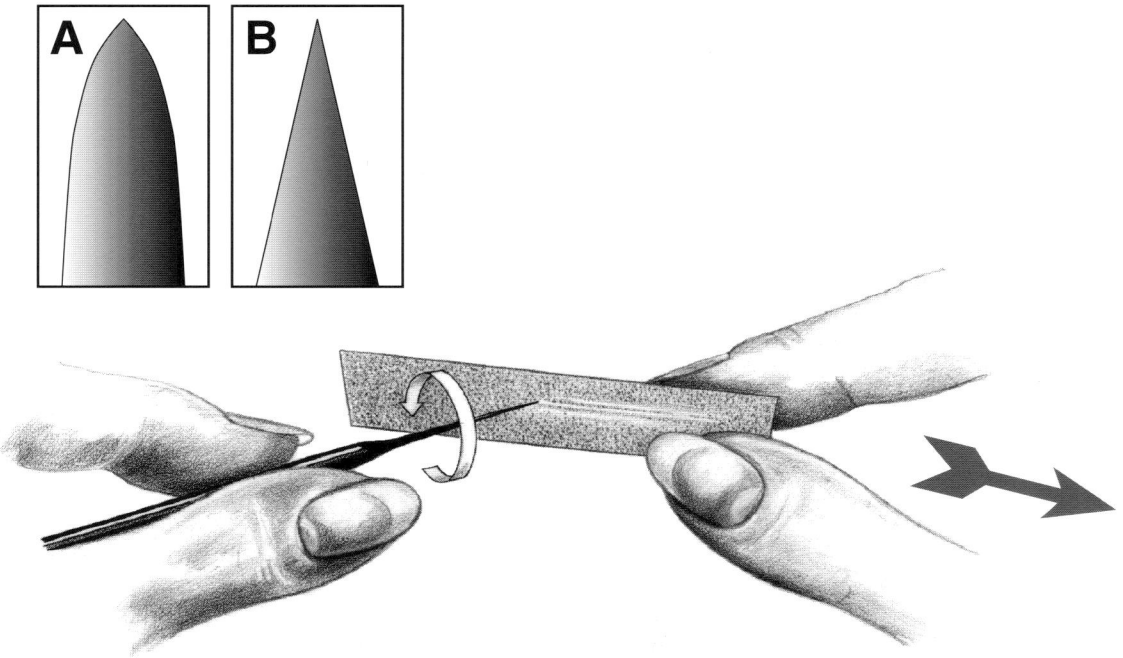

Figure 29
Sharpen the needle using fine sandpaper. Drag the sandpaper along the tip of the needle and twirl the needle to ensure a uniform point. A) Common needle tip for electrolysis. B) Sharp needle is best for telangiectasia removal—needle must pierce the skin.

PROCEDURE FOR REMOVING TELANGIECTASES USING THE BLEND

Absolutely none of the following treatment recommendations for telangiectasia removal should be attempted without authorization of a medical doctor. Only a licensed physician can judge which vessels can and cannot be treated!

> You will find that a sharp .003 inch diameter needle will work for most telangiectasia removal with the blend. Use the finer .002 diameter for very tiny vessels; use the thicker .004 diameter for larger vessels.

Select a short needle with a small diameter (.002, .003 or .004 inch diameter is preferable).[14] The needle must have a sharp point to facilitate easy skin penetration. Needles of correct size with sharp points are now being made specifically for telangiectasia removal with the blend—check with your medical equipment supplier.

Standard .002, .003 and .004 diameter electrolysis needles are also usable; however their tips are usually too rounded. If necessary, you may sharpen the tip using fine 600 to 1000 grit carborundum sandpaper and polish with crocus cloth (Figure 29). Certainly, you must sterilize the sharpened electrolysis needle before use!

---

14) Use a small diameter needle for very tiny blood vessels. A larger diameter needle may be used for larger vessels.

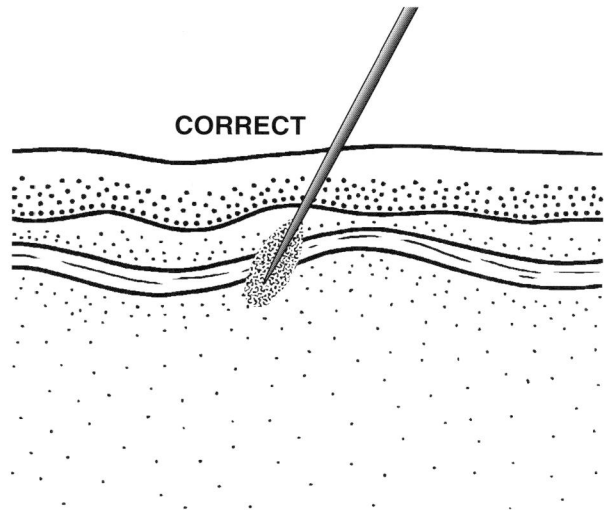

Figure 30
Do not insert the needle beyond the capillary. Coagulation takes place only at the tip of the needle. A "deep" insertion will not coagulate the vessel.

Since a shallow insertion is being made into a wet blood vessel, only low to medium HF levels are needed. The HF produces greater heat when insertions are shallow and tissue is wet.[15] (Low HF is known as "face-technic" by electrologists.)

The high frequency current is most intense at the tip of the needle. Therefore, only the tip of the needle is inserted into the blood vessel. Do not insert deeper than the vessel, because coagulation will take place below the vessel. In most cases the insertion will be less than 1 millimeter in depth (Figure 30).

The DC is set between 0.2 and 0.4 milliamperes. Both HF and DC levels are determined by the size of the vessel being treated. Small vessels require low current levels; larger vessels require higher current. There are several distinct treatment modifications.

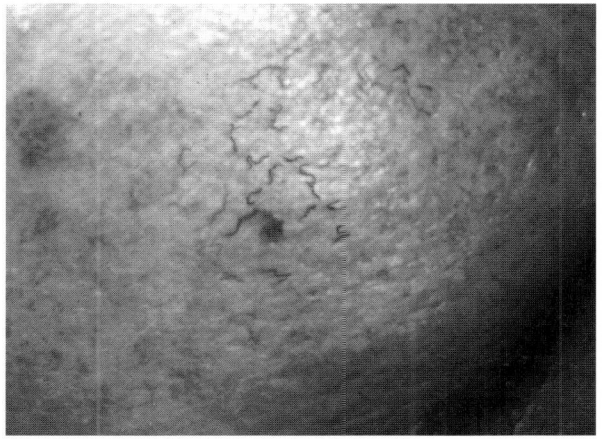

Figure 31
Common telangiectases seen on the cheek of a 54-year-old woman of Scottish heritage.

---

15) Michael Bono, *Real World Electrology: The Blend Method*, 1995, pages 78-79.

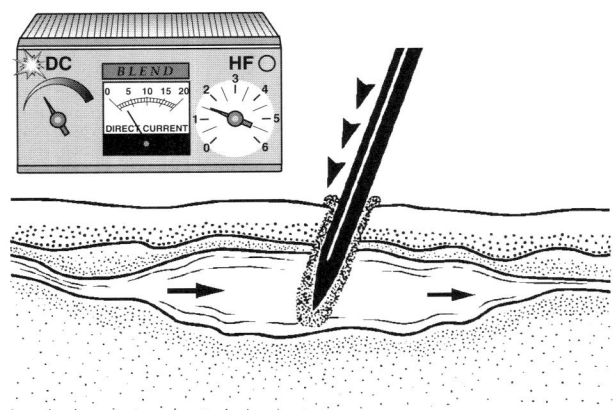

Figure 32
Insert with DC on.

Figure 33
Coagulate vessel with both DC and HF.

BASIC PROCEDURE

Set the HF power at low "face-technic" level (40 to 45 volts—check your manual for the correct HF setting). Set the DC level at 0.2 or 0.3 milliamperes (check your manual to set the DC meter). With the DC on and HF off, insert the needle into the lumen (center) of the capillary.[16] Keep the DC on during the insertion, to allow easy penetration into the vessel through the epidermis (Figure 32).

As soon as the needle has entered the vessel, apply the HF current with the DC remaining on. Notice that the capillary segment turns white as it quickly coagulates. Continue both currents until there is no more visible capillary coagulation. In most cases, coagulation takes place in 2 to 3 seconds (Figure 33). More time may be necessary for larger vessels. Very minute vessels require only 1 second of HF.

The coagulated segment is the clot that blocks the flow of blood in the capillary. Maximum segment coagulation is achieved when no more whitening takes place (usually less than 3 seconds). When maximum coagulation has taken place, stop the HF but continue to apply the DC.

The needle is now in the middle of a coagulated dehydrated clot and could be stuck to the clot. Take care removing the needle so that the clot is not pulled out with the needle.

"Face-technic" is a term used only in electrology—meaning both currents are on at the same time, and at low levels. For telangiectasia removal, DC is set at 0.2 to 0.4 milliamperes, and HF is set at 40 to 55 volts. To find the correct HF levels, refer to the manual for your machine. Certain machines are made specifically for telangiectasia removal. Again, you must refer to your manual for correct current settings.

---

16) Lumen: The open area of the vessel through which blood flows. The term "lumen" also describes similar open areas such as intestinal lumen, i.e., the open space in the intestine.

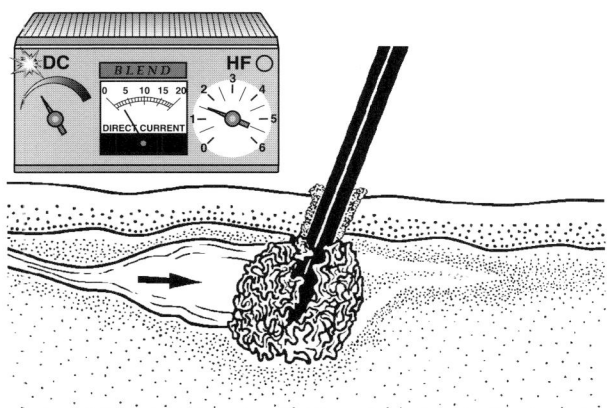

Figure 34
Clot is formed, but continue to apply the DC. HF is off.

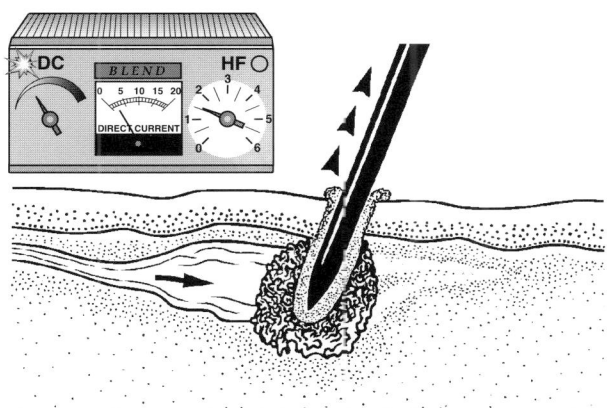

Figure 35
Continue to apply the DC. Some of the clot is dissolved to allow easy needle removal.

With the HF off, keep the DC on and the needle in place for about 1 to 2 seconds. The DC produces sodium hydroxide (lye) in the clot and dissolves enough coagulated matter to allow easy removal of the needle from the clot and skin (Figures 34 and 35).

Gently withdraw the needle from the coagulated clot (Figure 36). Observe the skin. It should not pull up or "tent" as the needle is being withdrawn. If the skin pulls up, the needle is sticking. Always allow the DC current to flow until the needle releases easily from the clot.

For nearly all telangiectasia removal with the blend, you will use current levels as just described under **basic procedure.**

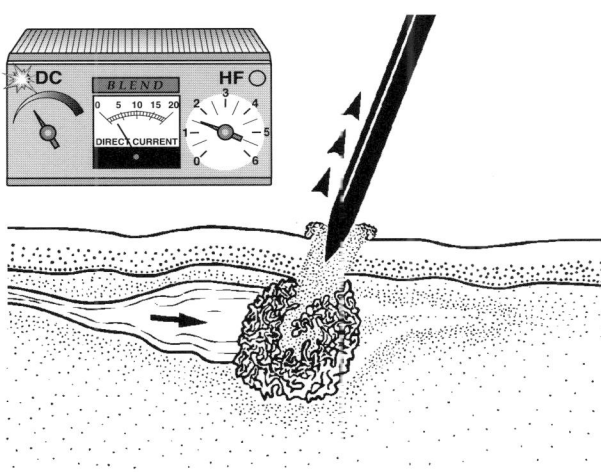

Figure 36
Continue to apply the DC as the needle is removed.

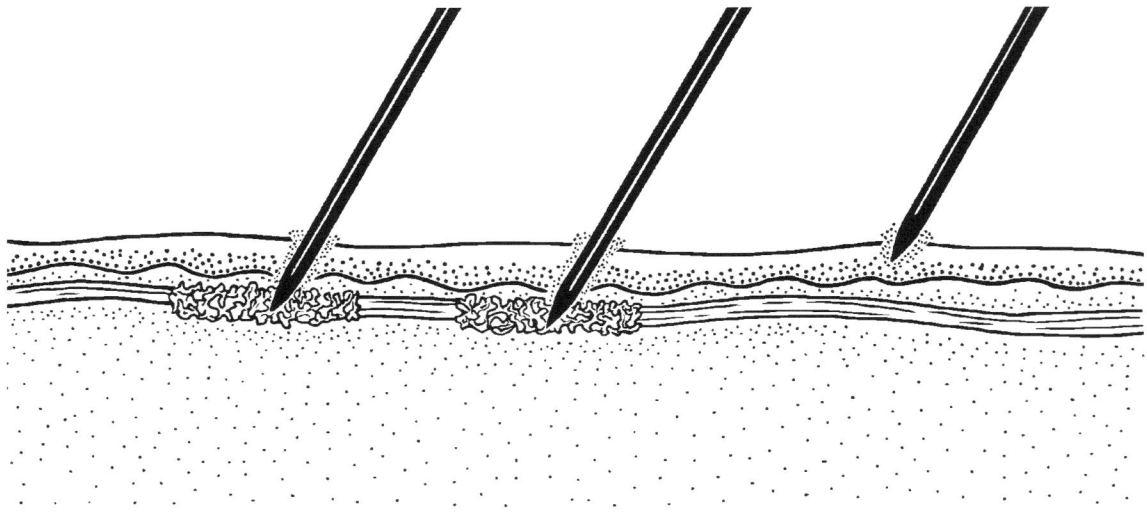

Figure 37
Properly space insertions to avoid overtreatment. Insertions are usually about 2 to 3 millimeters apart.

LINEAR VESSELS

If treating a linear vessel, attempt to coagulate its entire visible length. A linear vessel is usually not more than 1 centimeter long and may be straight, curving or tortuous (Figure 38). (Also, see photo 1, page 65.)

Note the approximate length of the coagulated clot (usually about 2 to 3 millimeters). The next insertion into the linear blood vessel should be about the same distance as the entire length of the clot itself. For example, if the first coagulated clot was 3 millimeters long, the next insertion will be at least 3 millimeters from that first insertion point (Figure 37). Insertions should not be too close together (under 2 millimeters) as this can cause overtreatment, excessive scabbing and scars.

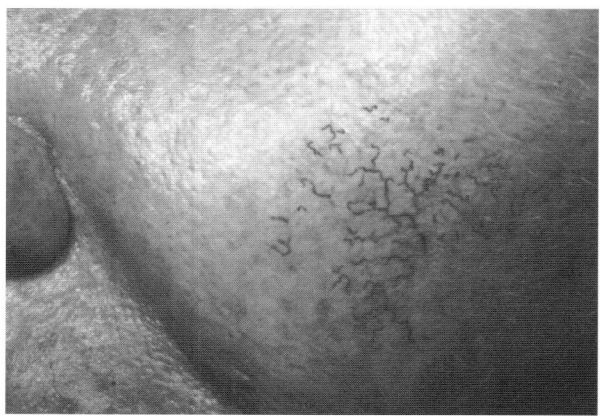

Figure 38
The majority of the telangiectases you will be treating are classified as linear telangiectasia. Even though the vessels may be tortuous or meandering, if you see discernible vessels with few radiating or clustering vessels, the telangiectases are linear.

MODIFIED BASIC PROCEDURE

When treating a large or fast-flowing telangiectasia, bubbles will sometimes form in the vessel (Figure 39). The DC produces these tiny hydrogen gas bubbles that flow rapidly downstream.

Unfortunately, DC bubbling usually prevents treatment success. Coagulation and vessel blockage is not accomplished because the bubbles agitate and carry away the coagulating blood as it forms. Should bubbling occur, modify the treatment procedure as follows.

First, insert the needle into the capillary using the **basic procedure** as just described (insert with the DC on and then apply the HF). Then, if bubbling appears, turn off the DC and continue to coagulate the blood vessel with the HF only (Figure 40). When coagulation has been achieved, turn the HF current off and turn the DC back on to remove the needle from the coagulated clot (Figure 41). Using the currents in this way is called the **modified basic procedure**.

Fast flowing blood also cools the needle by carrying off the HF heat. Because of this, coagulation can be prevented. If coagulation is unsuccessful using low HF power, compensate for the cooling-effect by slightly increasing the HF power. You may incrementally increase the HF until the vessel coagulates with no visible overtreatment.

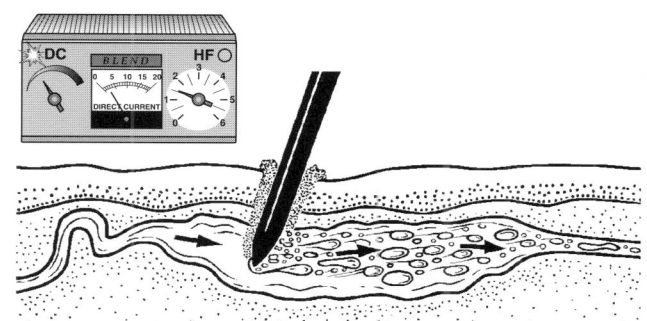

Figure 39
During insertion, bubbles are seen. DC only is on.

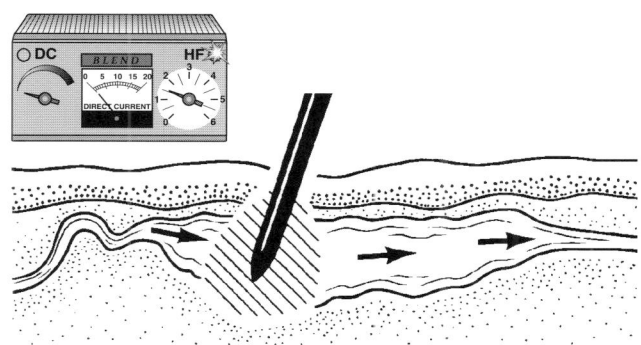

Figure 40
If bubbling continues, coagulate with HF only. DC off!

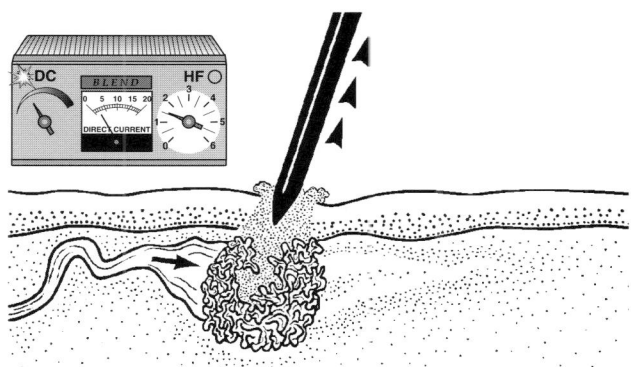

Figure 41
Remove needle with DC only. HF off.

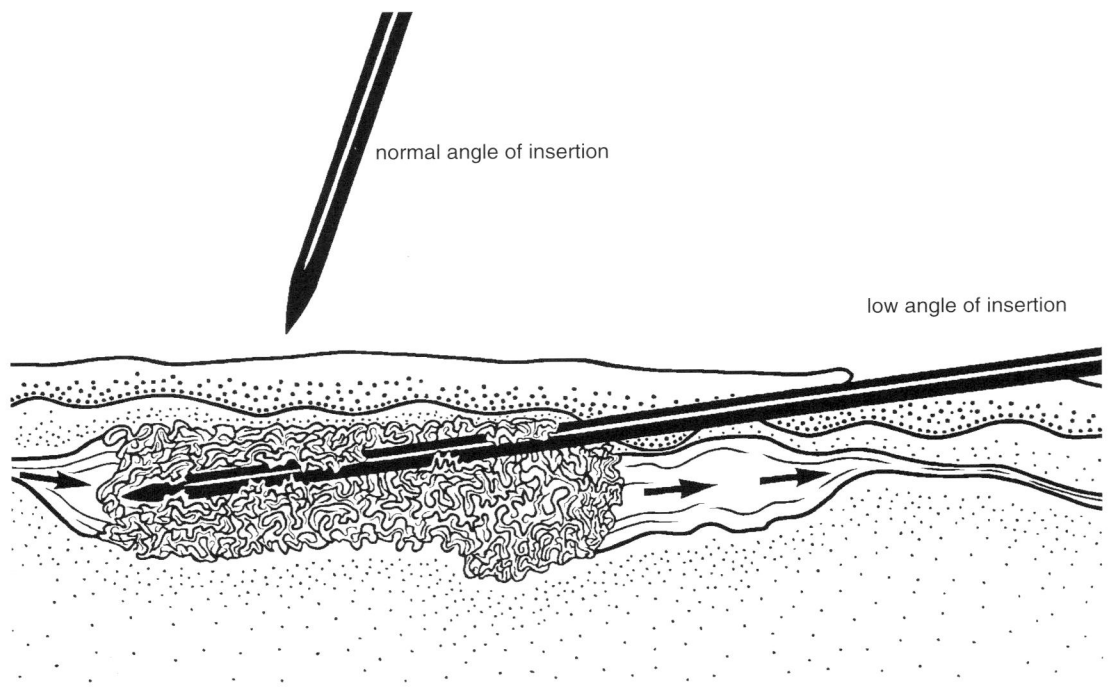

Figure 42
Insert at a low angle to the skin if the blood vessel is large or fast flowing. In this way, the needle contacts a greater area of the vessel, coagulates more matter, and ensures treatment success.

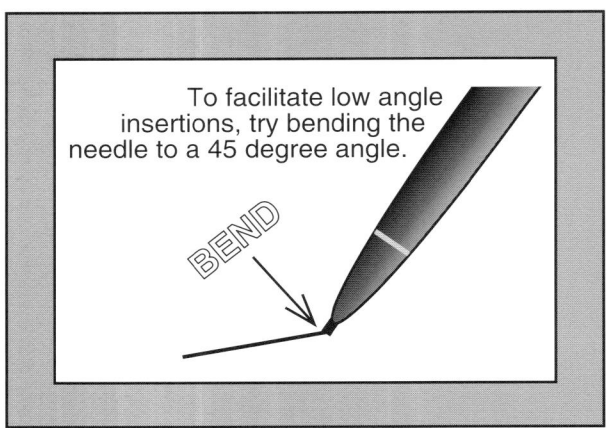

To facilitate low angle insertions, try bending the needle to a 45 degree angle.

If a large capillary will not coagulate easily, try inserting the needle at a low angle (Figure 42). Doing this causes the needle to contact more of the capillary thereby coagulating a larger segment. Be sure to allow the DC current to completely loosen the needle from the clot before withdrawing the needle.

OVERTREATMENT

An overtreatment from telangiectasia removal does not look like overtreatment from an electrolysis treatment. In electrolysis, you look for any changes in the skin's surface that indicates that HF has reached the top layers of the skin. If the top layers are affected, you can assume that significant damage has taken place in the deep dermis, and visible scar tissue may form.

In telangiectasia removal, the currents coagulate the skin's surface—the epidermis is always coagulated. However, because only the superficial dermis is affected, scar tissue will not form. The epidermis rapidly regenerates following treatment, and leaves no mark whatsoever.

Each needle insertion should produce a tiny spot of coagulated skin. The needle should release easily from the spot; there should be no sticking. Each spot should be isolated, that is, the spots should not merge together to form a single coagulated mass. If the skin seems to pull together, "tent" or stick to the needle, your HF is probably too high, and you are overtreating the skin. If the coagulated spots merge together, you are overtreating the skin—your insertions are too close together.

EXTENSIVE CONDITION

Some patients have a widespread condition with hundreds of fine telangiectases which gives the skin an overall red appearance. Such a condition often exists on the upper chest or neck (Figure 43). In this case significant speed can be achieved by slightly altering the treatment procedure.

First, remove one vessel using the **basic procedure**, but this time withdraw the needle from the skin with both the DC and HF currents on. Look to see that no bubbling takes place in the capillaries, the needle releases easily from the coagulated clot and there is no overtreatment. If you observe all these conditions, continue using the following procedure.

Keep both the DC and HF currents on at all times, and insert the needle from capillary segment to segment. Insert the needle randomly, it is neither possible to determine the direction of blood flow in the vessels nor is it important to the success of the treatment. A slight increase in the DC power is sometimes necessary to allow easy needle insertion and removal. (In most cases, a DC increase of 0.1 to 0.2 milliamperes is all that is necessary.)

Attempt to coagulate the entire length of any linear vessels within the treatment area. If bubbling is seen in the blood vessel, return to using the **modified basic procedure**. Bubbling indicates a fast-flowing vessel that must be coagulated using HF alone; the DC must be off to allow this.

If the skin is very moist, inserting and removing the needle with the HF on can cause overtreatment. Look for this potential problem as the first few insertions are made. Should a problem be seen, return to using the **basic procedure**: insert with DC, coagulate with DC and HF—then remove the needle from the skin with DC.

Figure 43
Upper chest of a 42-year-old woman with an "extensive condition." Treatment speed can be achieved by keeping the DC and HF currents on at all times during the procedure.

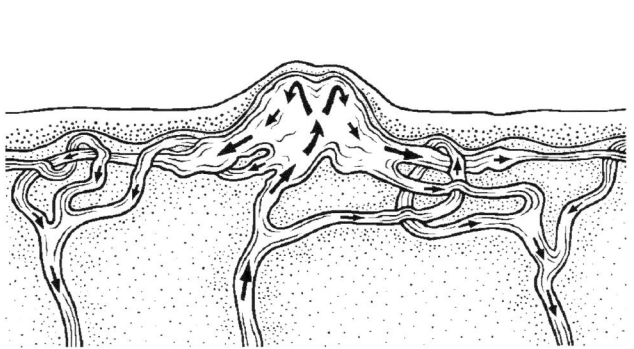

Figure 44
A spider telangiectasia often protrudes above the skin. Arrows show the upward direction of blood flow that causes this elevation.

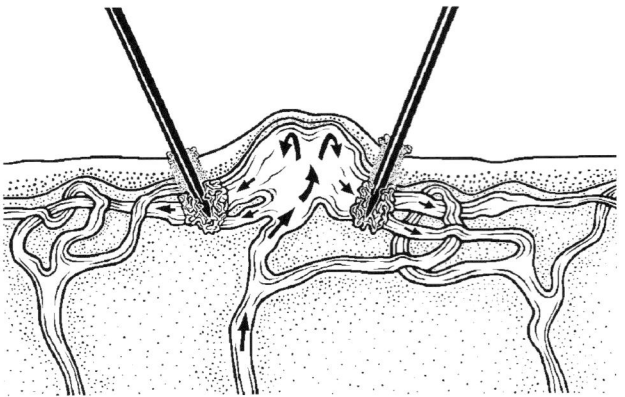

Figure 45
First, coagulate a few larger connecting capillaries.

SPIDER TELANGIECTASIA

A spider telangiectasia has a tiny arteriole or capillary at the center with radiating capillaries. The center looks like a spider's body and the radiating capillaries look like the legs. In most cases, blood is flowing up from the center and out into the capillaries; the center is often slightly raised (Figures 44 and 48). The upward flow of blood causes the spider center to be elevated. In most cases, the raised skin becomes level when the vessel is eliminated (Figure 47).

> If you are unsure of the direction of blood flow in a telangiectasia, try the following. Gently press down on the vessel with your finger tip. This pressure momentarily stops the blood flow, and the vessel empties of blood. When you release your finger, blood then flows back into the vessel, and you can see the actual direction of the blood.

Slight finger pressure applied to the center of the spider can interrupt blood flow and cause the connecting capillaries to disappear momentarily. (In some cases blood is flowing in the opposite direction—from the capillaries into the center of the spider—although this is less common.)

A raised stellate lesion is sometimes called a spider angioma: *angi*, meaning blood or receptacle and *oma*, meaning a swelling or enlargement. (The term "spider telangiectasia" is more commonly used.)

Removal of a spider telangiectasia is performed as follows. First, coagulate several larger connecting capillaries (Figure 45). Then, insert the needle directly into the center of the spider and treat. Insert to a depth of 2 millimeters or more (Figure 46).

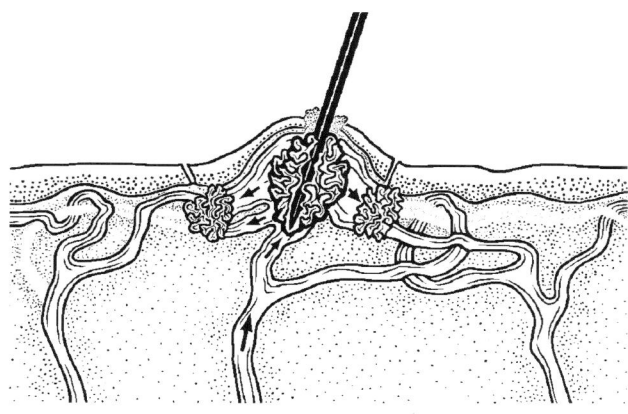

Figure 46
During the same treatment, coagulate the center of the spider.

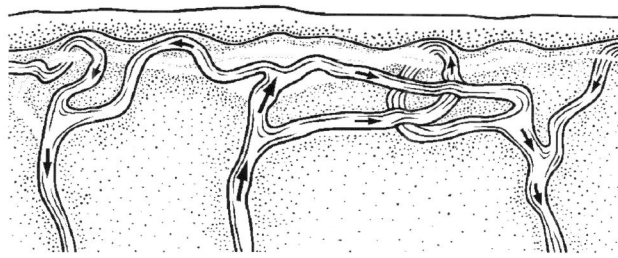

Figure 47
When the spider telangiectasia is successfully removed, the skin level usually returns to normal.

Use the **basic procedure**, which is: insert with DC, coagulate with both DC and HF and remove the needle with DC only. It may be necessary to increase both currents and allow more time for coagulation to take place than with smaller telangiectases. If bubbling takes place, use the **modified basic procedure**, as explained earlier (see page 41).

Attempt to coagulate the entire spider center leaving no blood flow. It might be necessary to insert a few times at different angles to achieve coagulation. Do not overtreat the skin, however, by making too many insertions; usually two are sufficient. If total coagulation is not achieved, the vessel can be treated at another time. A scab of 1 to 2 millimeters in diameter will form over the treated spider center.

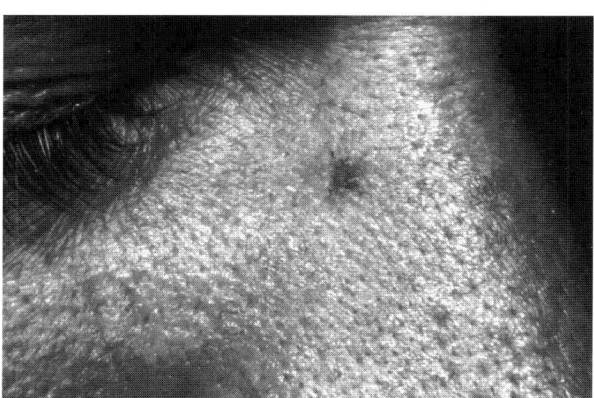

Figure 48
Common spider telangiectasia seen on the nose. Patient thought the spider telangiectasia was caused by heavy eyeglasses resting on the bridge of his nose.

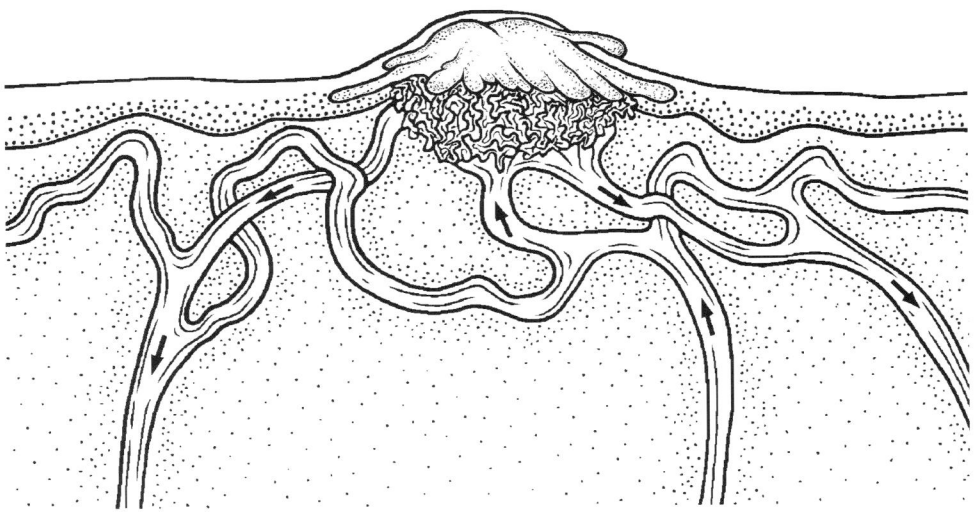

Figure 49
Blood sometimes leaks and "pools" under the skin, even though the telangiectasia has been properly treated.

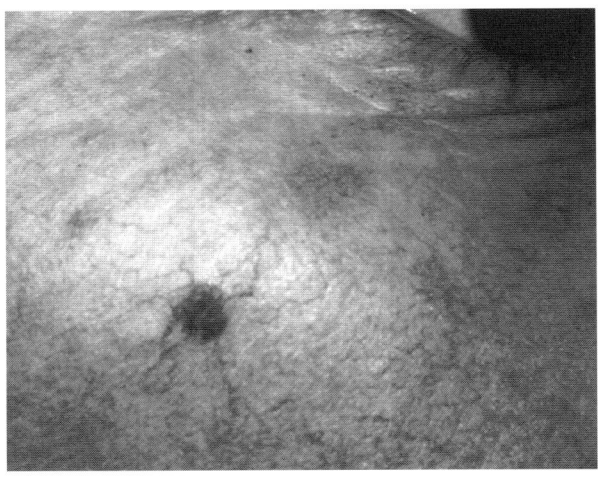

Figure 50
Very large spider seen on the cheek. Such a large spider must not be completely coagulated in one session. Treat half or one-third, then wait several weeks to complete the job. Completion may take 3 or 4 treatments. (Photo shows lesion before treatment.)

Occasionally, a properly treated spider telangiectasia fills with blood immediately after the treatment.[17] In this case, blood seeps out of the treated vessel and forms a pool under the epidermis (Figure 49). Blood may leak out onto the skin surface through the insertion site. The appearance of this blood does not necessarily indicate treatment failure. (See photo 11, page 67.)

Do not treat the leaking blood, simply allow it to pool under the skin. Attempting to coagulate the leak usually causes overtreatment. Instruct the patient to expect a dark red scab of about 2 millimeters in diameter to form. The patient may return for treatment (in two to three weeks) when the scab has sloughed off and if the telangiectasia is still visible.

---

17) A linear telangiectasia may also leak after treatment. However, larger vessels with greater blood flow, such as the typical spider telangiectasia, have a greater leakage potential than the average linear vessel.

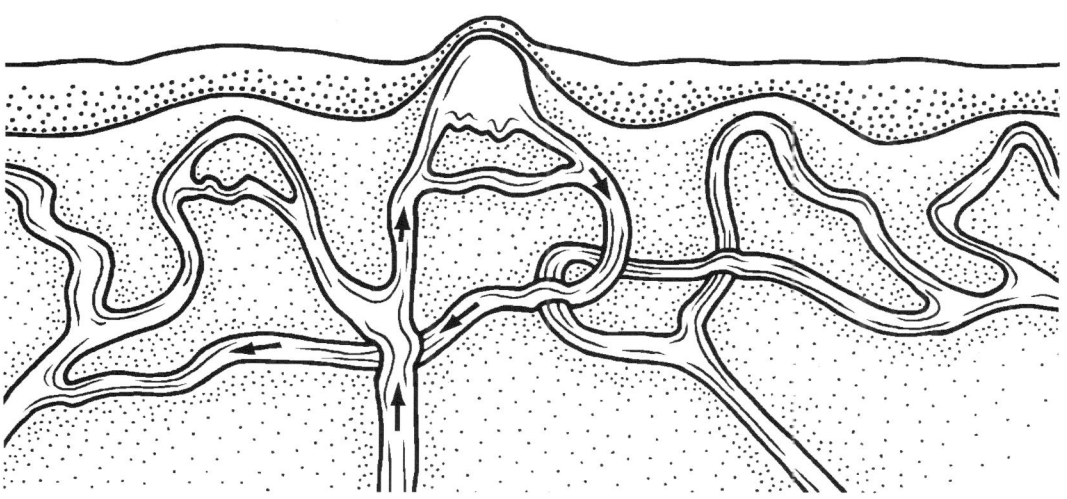

Figure 51
The dot-telangiectasia appears as a simple red dot (or several dots) with no radiating capillaries.

DOT-TELANGIECTASIA

Some telangiectases are simple red dots with no adjoining capillaries. This so-called "dot-telangiectasia" is typically level to the skin surface but is sometimes slightly raised (Figures 51 and 52). Most dot-telangiectases are less than 2 millimeters in diameter. These lesions are often called "blood spots" and are seen on the lips, nose and under the eyes. Improper comedo extraction (squeezing the skin) can sometimes cause a dot-telangiectasia. (See photo 3, page 65.)

Treat a dot-telangiectasia the same as a spider telangiectasia center. Treat a dot only if it is small, bright red and nearly level to the skin. If it is large, dark red or blue and is more than 2 millimeters above the skin surface, it might be a large protruding vein and should not be treated. Ask the cooperating physician for reevaluation if you are uncertain.

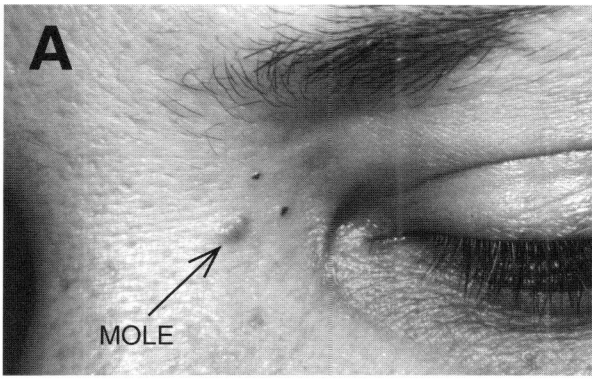

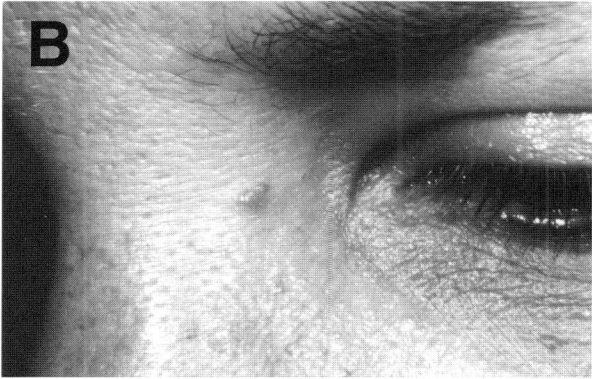

Figure 52
Dot-telangiectases are often seen on the nose, lips and eyelids. A) 37-year-old woman with two dot-telangiectases on the eyelid area. B) Eight weeks after one treatment. Usually, such dot-telangiectasia require only one treatment with the blend method.

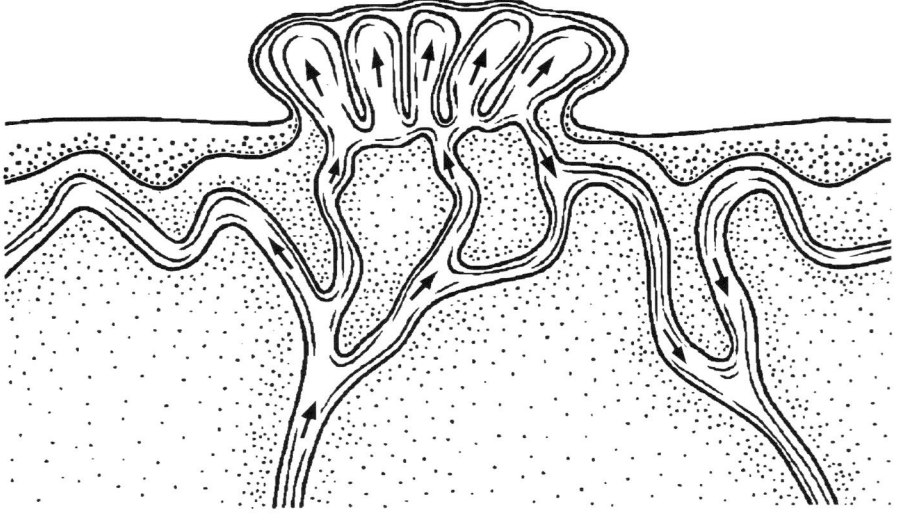

Figure 53
"**Blood mole**." Some telangiectases look like large red moles and protrude above the skin line. Such lesions are common on the trunk, but rare on the face and legs.

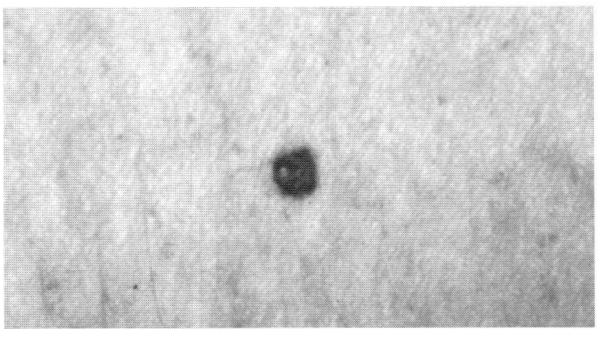

Figure 54
Common "**blood mole**" seen on the abdomen of a 26-year-old Caucasian woman, before treatment.

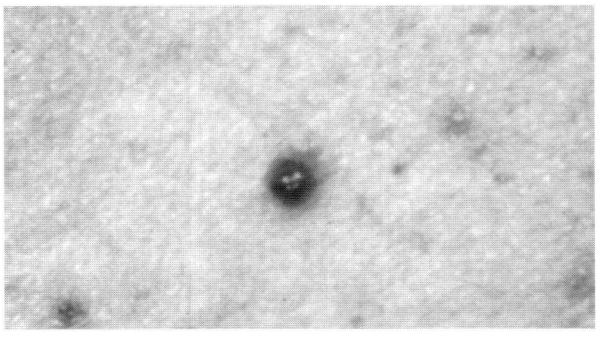

Figure 55
"**Blood mole**" on the back of a 30-year-old man. Above photo shows blood mole immediately after treatment—notice the 4 white coagulated dots. The entire lesion was not coagulated at one time; subsequent treatment removed the remaining lesion.

"BLOOD MOLE" ANGIOMA

A telangiectasia may protrude several millimeters above the skin surface and resemble a nevus (mole). This type of angioma is commonly called a "blood mole" and is typically found on the chest and back. It may be consequential that this category of lesion is often seen below the breast where the bra rubs the skin.

The blood mole angioma is a cluster of dilated capillaries protruding up, just under the epidermis. The skin covering the lesion is extremely thin. The blood mole angioma is often more than 4 millimeters in diameter (Figure 53, 54 and 55).

Treat the blood mole using the **basic procedure**. Coagulate each individual vessel in the cluster. In most cases (if under 3 millimeters in diameter) the lesion can be entirely coagulated at one time. However, if the angioma is larger than 4 millimeters in diameter, it should be treated in two or more separate appointments. Treat only one-half or one-third at each appointment (Figure 55).

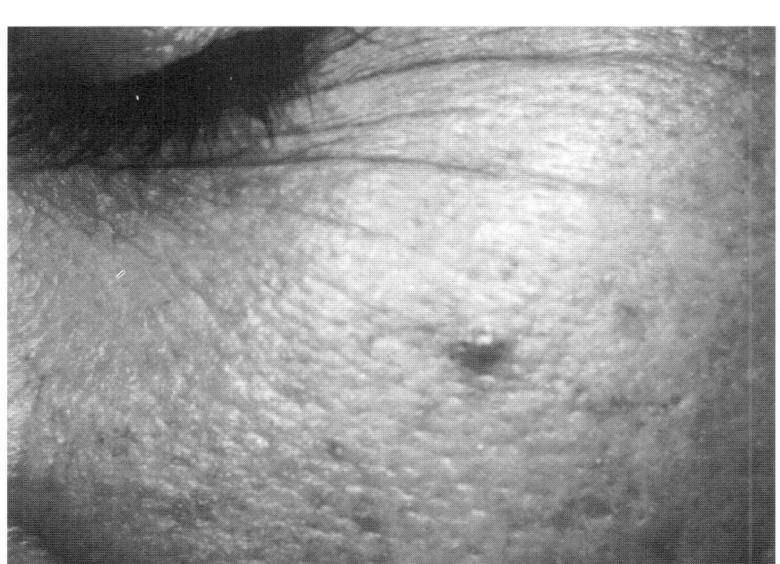

Figure 56
Protruding telangiectasia: "Blood moles" on the body look like they are made up of tiny blood droplets. Blood moles on the face usually consist of one or more linear blood vessels. (On the face, treat the individual vessels. On the body, you may treat in a "random manner" if individual vessels are not discernible.)

Because the "blood mole" angioma protrudes significantly above the skin surface, coagulation takes place well above the dermis. Although a rather large crust of 3 to 4 millimeters forms after the treatment, the skin almost never exhibits marks or scars after crusts have sloughed off.

RED BLOTCHY AREAS

A patient may have a red blotch with no visible blood vessels. In this case capillaries are so small they cannot be individually seen and therefore cannot be separately coagulated. Red blotches are commonly seen on the neck and upper chest. Although you cannot see distinct vessels, blotches can be successfully treated (Figure 57).

Randomly treat the blotchy area, leaving at least 3 millimeters between insertions. After several treatments, small invisible vessels are cleared away and larger well-defined vessels become visible. Treat these visible capillaries using the **basic procedure**. If blood vessels do not become individually visible, continue treating the blotchy lesion in a random manner.

Figure 57
Blotches are seen on the neck and upper chest. Skin is thin in this area and is frequently exposed to sunlight (open-necked shirts). This overexposure to sunlight probably causes the condition.

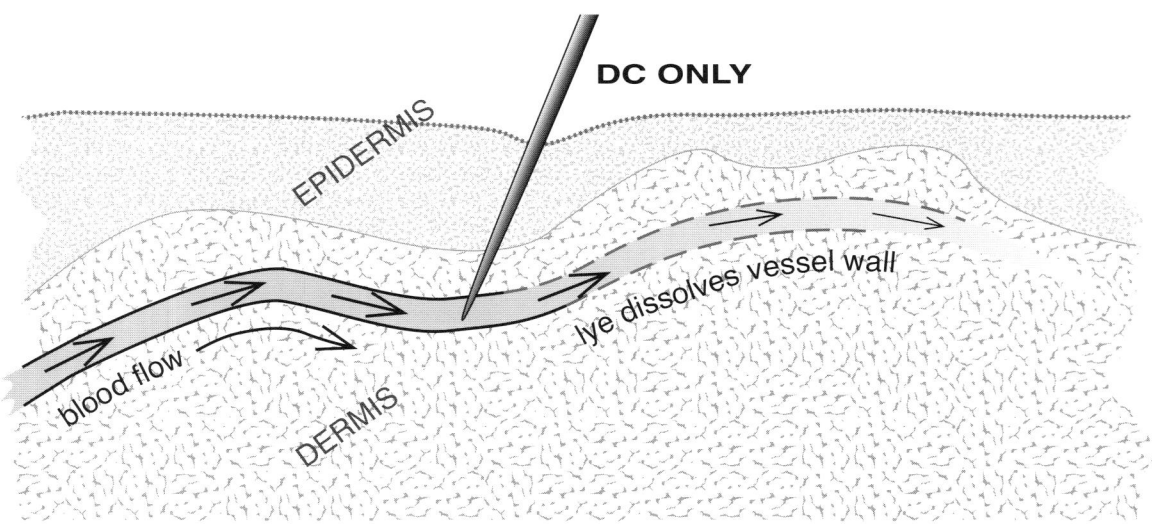

Figure 58
DC only may be used for minuscule telangiectases. DC produces lye that travels downstream in the vessel. The small amount of lye is sufficient to dissolve the entire downstream portion of the vessel wall itself. The destroyed vessel clots, and is then absorbed by the body.

DC ONLY PROCEDURE

Very small telangiectases lie close to the skin's surface (just under the epidermis). These capillaries are often seen on the bridge of the nose and under the eye. Very small telangiectases can be removed using the DC current alone (Figure 58). By not using the HF, damage to the skin is minimized and post-treatment crusts are microscopic.

Use 0.2 or 0.3 milliamperes of DC current. Insert into the capillary with the DC on. Sodium hydroxide (lye), is produced and dissolves the entire downstream portion of the vessel wall.

This procedure is only effective in treating minuscule blood vessels. If the vessel is average in size or fast-flowing, the HF current must be used to coagulate the vessel. Try experimenting: begin with DC only, and if the vessel does not dissolve add the HF to achieve vessel coagulation.

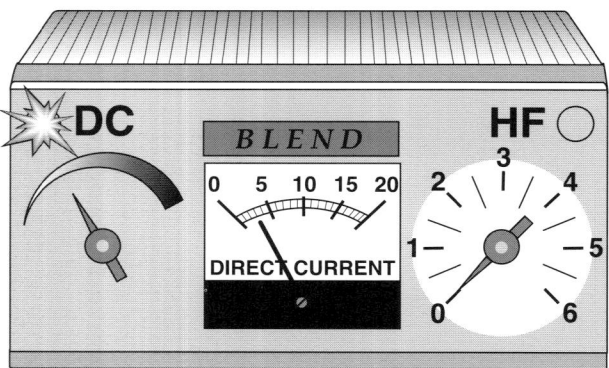

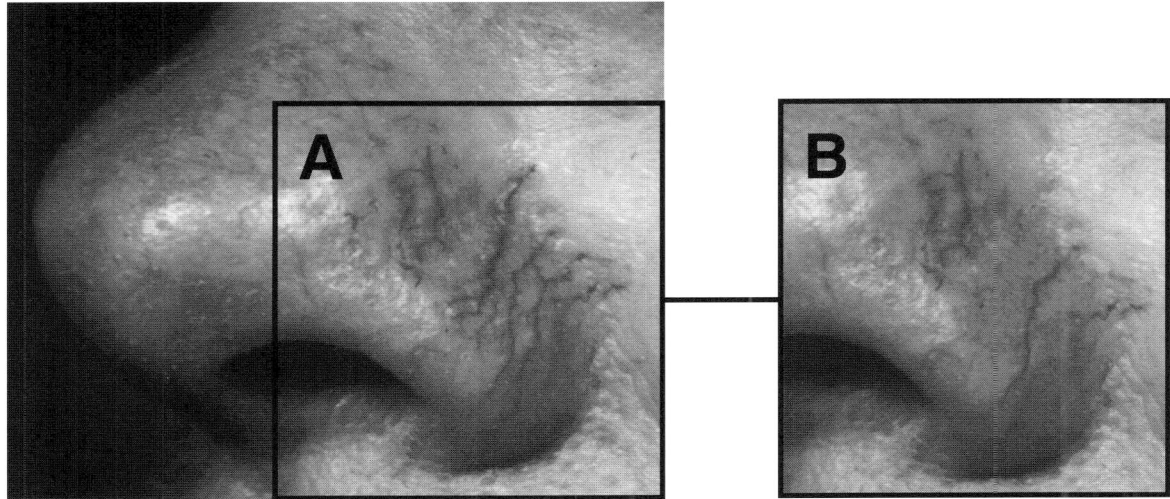

Figure 59
Common telangiectases on the nose: A) Before treatment. B) Five weeks after the first treatment. Notice that not all vessels were removed, instead only about half were removed. Removing telangiectases in sequential stages minimizes overtreatment and encourages treatment success.

REVIEW

There are many different types of telangiectases; each requires a slightly different treatment approach. To review: (1) most common linear telangiectases are treated using the **basic procedure** with both HF and DC. (2) Larger fast-flowing blood vessels are coagulated using the **modified basic procedure**, HF only—DC off. (3) An extensive condition can be treated by keeping the HF and DC on continuously. (4) Spider telangiectases require a deeper insertion, more time and higher HF current. (5) Dot-like markings are treated like the spider telangiectasia center. (6) Red blotches can be treated randomly to clarify the area. (7) Very small telangiectases can be treated with DC alone.

While each type of telangiectasia requires a slightly different approach, several recommendations can be made for treating them all. The following are these suggestions.

> Telangiectasia removal with the blend uses several variations. Typically, you will use several technique variations during one session. For example you might find a linear vessel, a dot-telangiectasia, and a spider, all in the same area. You may even encounter a tiny vessel requiring only DC.

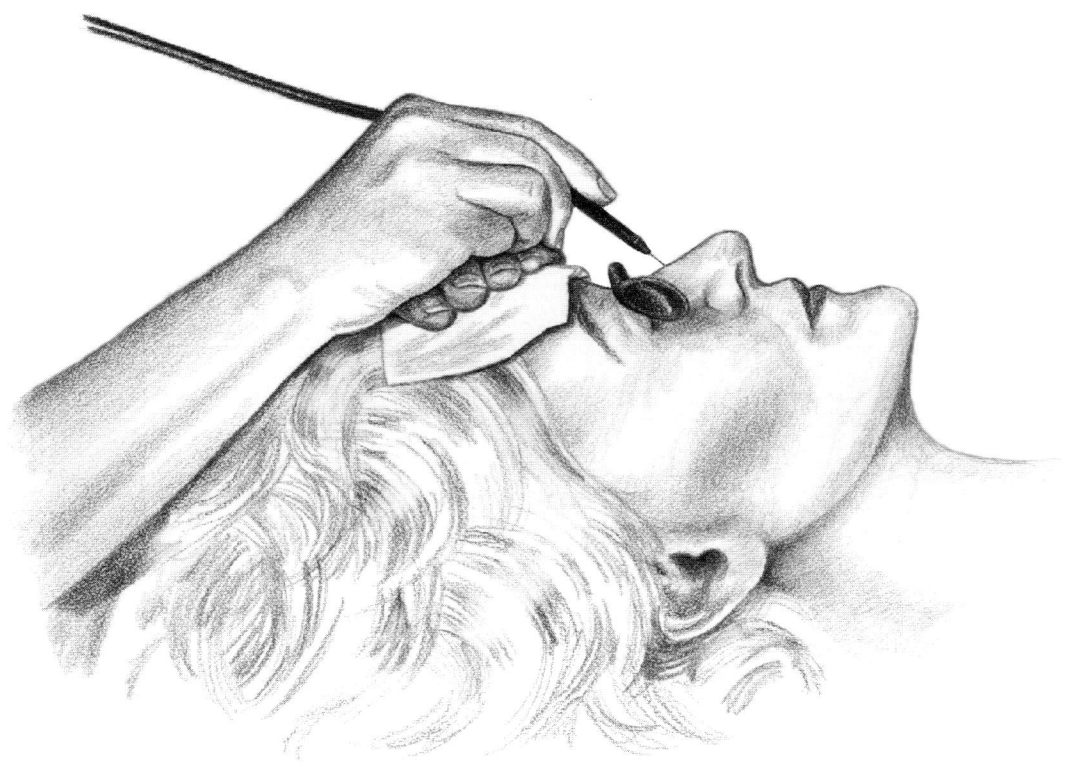

Figure 60
Illustration shows positioning for treating the nose. Begin from the center of the face. Stabilize your working hand by resting it gently on your opposite hand. Do not stretch the skin. Start at the bridge and work out toward the corners.

INCREASED BLOOD FLOW AND STRETCHING THE SKIN

Advise the patient to refrain from any activity that increases blood flow to the skin at least one hour before treatment. Sauna, exercise, sunbathing, and similar activities must be avoided. Increased blood flow makes vessel coagulation more difficult.

Do not stretch or press on the skin during treatment. Pressing or stretching can open the coagulated segments. To avoid accidentally pressing on the treated tissue, do not rework the treated area. Always work from the center of the face outwardly: start from the nose and work toward the cheeks (Figures 60 and 61).

Figure 61
Treatment on the nose and cheeks. Begin treatment from the center of the face; work outwardly. Rest your working hand on the opposite hand. Gently press the skin toward the center of the face, to ensure you do not stretch the skin (see arrows).

LOCAL ANESTHETIC

Injectable local anesthetic such as Xylocaine® should not be used.[18] Anesthetic constricts capillaries, rendering them difficult to see and thus treat. The pressure of the liquid anesthetic under the skin also temporarily obliterates small vessels.

Removing telangiectases with the blend is not painful. But, if a local anesthetic is necessary the doctor should use a formulation without epinephrine in order to diminish the effect of vessel constriction. After injection, a short waiting period will give the obliterated vessels a chance to appear once more.

> The topically applied anesthetic cream called EMLA is suitable for telangiectasia removal. Apply the cream thickly to the skin one hour before treatment. Most patients report about 50 percent pain reduction with EMLA. The anesthetic action does not significantly conceal the vessels.

---

18) Xylocaine is a trade name of the Astra Company. It is a formulation of water, lidocaine hydrochloride and epinephrine. (Xylocaine is also available without epinephrine.)

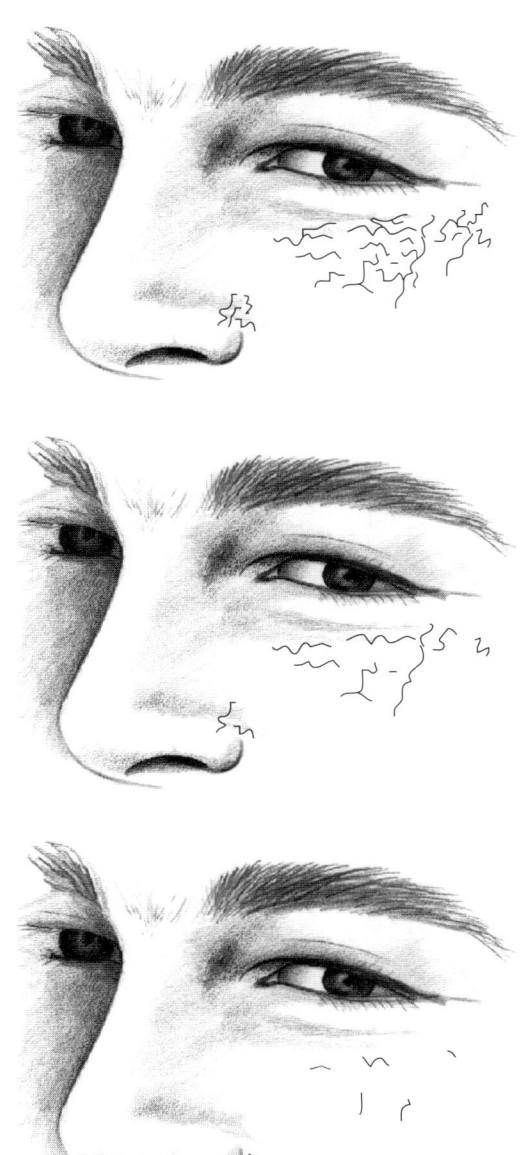

Figure 62
If the telangiectases are especially numerous or concentrated, remove in progressive stages. Always allow at least two to three weeks between each treatment.

ASPIRIN AND ICE

Aspirin must not be used to subdue pain prior to treatment. Aspirin acts as an anticoagulant and increases the likelihood of the vessel leaking following treatment. The area should not be iced prior to treatment. Icing constricts blood vessels, rendering them difficult to see.

TREATMENT TIME FOR ALL PATIENTS

Start each patient with a preliminary treatment. If the patient has only one lesion, try one or two insertions. If the case is extensive, such as the entire upper chest, a five minute treatment is sufficient. Observe the patient's healing ability by waiting at least two weeks until the next treatment.

After the two week period, the skin should appear normal and some of the lesions should be removed. Treatment time can now increase to the recommended maximum of five to ten minutes (one treatment every two weeks).[19] In this way the telangiectases can be removed in progressive stages (Figure 62).

---

19) Treatment time varies greatly: 10 minutes is for an extensive case such as an entire upper chest. Most areas, however, require only a 2 to 5 minute treatment.

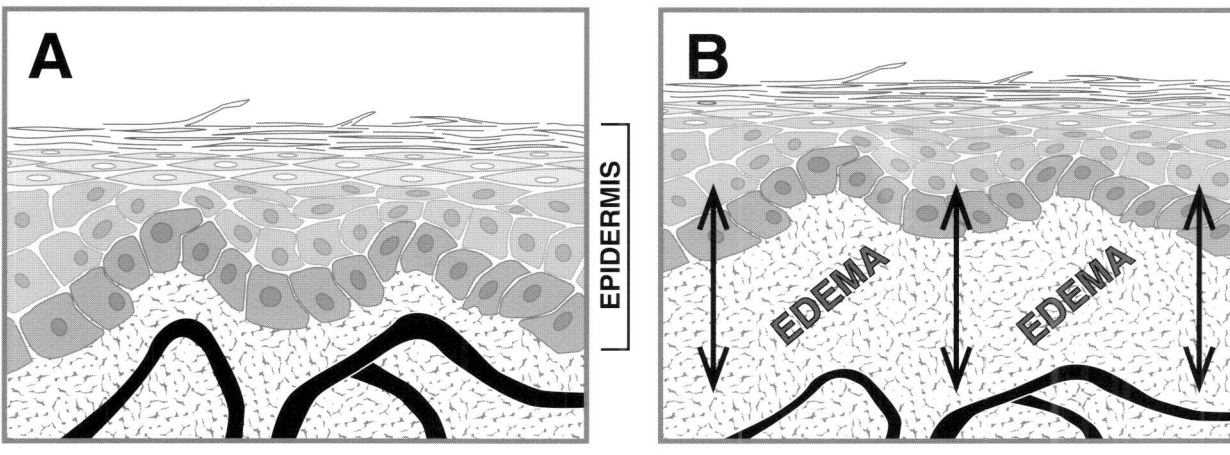

Figure 63
A) In untreated tissue, telangiectases are visible through the transparent epidermis. B) Edema from treatment expands the dermis, stretches the epidermis and flattens adjacent telangiectases. Thus, edema often makes seeing nearby telangiectases impossible. You may treat these at a subsequent appointment.

OVERHEATED SKIN

The immediate area becomes quickly swollen from the treatment itself. This swelling, or edema, is normal and will not effect treatment success. However, because of this swelling, many telangiectases in the surrounding area become impossible to see (Figure 63). Simply treat these vessels at another appointment. It usually takes three separate treatments to achieve the desired results in one specific area.

Too long a treatment can cause the entire area to become hot. When the skin becomes hot, blood flow increases which can cause the coagulated segments to open. If the skin becomes hot and red from the treatment itself, discontinue treatment until the next appointment. Always allow the inflammation to completely resolve.

CONCENTRATING INSERTIONS

Do not concentrate the insertions. Blood vessels supply life-giving nutrients to the cells. Severe overtreatment can reduce the blood supply to the skin. Necrosis (skin death) can result, leaving permanent scars.

Place insertions at least 2 to 3 millimeters apart. Treat an area no larger than 1.5 square centimeters. Move to another location, if needed, but separate the areas by at least 2 centimeters or more.

> Most patients undergoing telangiectasia removal become aggressive with treatment and want you to remove as much as possible in one session. Regardless of the patients' demands, you must not exceed safe treatment limitations. If the area begins to get hot and swollen, you must stop.

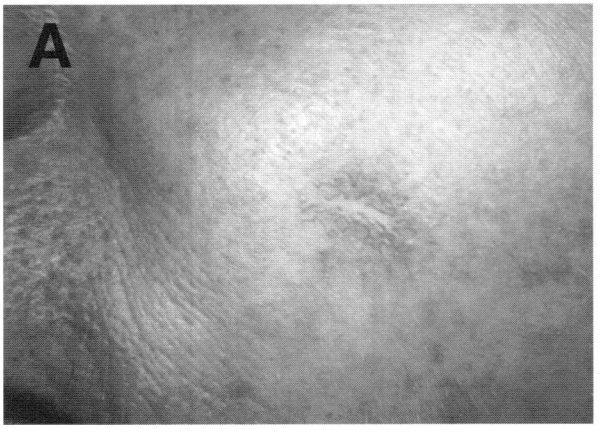 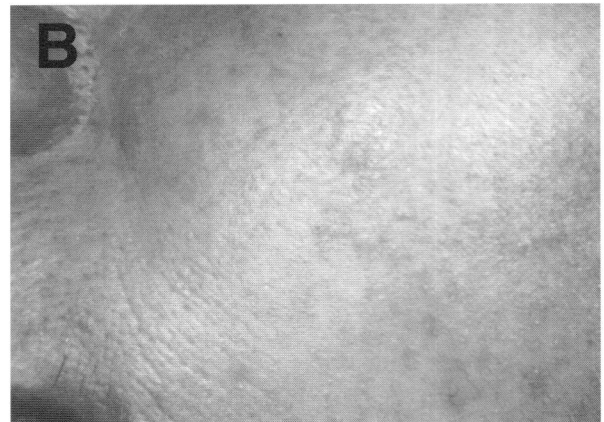

Figure 64
A) Hyfrecator was used to remove telangiectases from the cheek of a 46-year-old woman. The treatment resulted in highly visible scar tissue with surrounding remaining telangiectases.
B) Treatment with blend removed surrounding vessels, and made the scar less noticeable.

CATAPHORESIS

If the skin appears traumatized, cataphoresis is recommended. Cataphoresis decreases swelling and helps contract the treated blood vessels. Use between 0.5 and 1.0 milliamperes of DC current. Be careful not to press down on the skin with the cataphoresis roller. Too much pressure from the roller can open the coagulated segments. Place a cotton pad saturated with witch hazel[20] (or tap water) on the treated area and roll the cataphoresis electrode gently over the pad (Figure 65).

CRUST FORMATION

Tiny crusts will form 1 to 3 days after treatment. Crusts are usually less than 0.5 millimeter in diameter and slough off in seven to fourteen days. Large vessels require more current thus crusts will be larger, about 1.0 millimeter, and last up to two weeks. Tell the patient to expect these crusts and not to remove them forcibly.

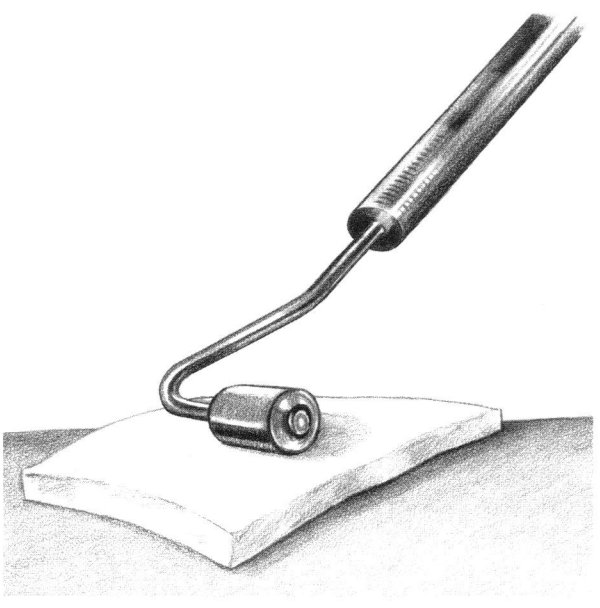

Figure 65
Use cataphoresis if skin appears overly inflamed. Place cotton soaked with witch hazel on the skin and gently roll over the area.

---

20) Witch Hazel is a weak solution of alcohol, water and herbs. It is a mild astringent and soothes the skin.

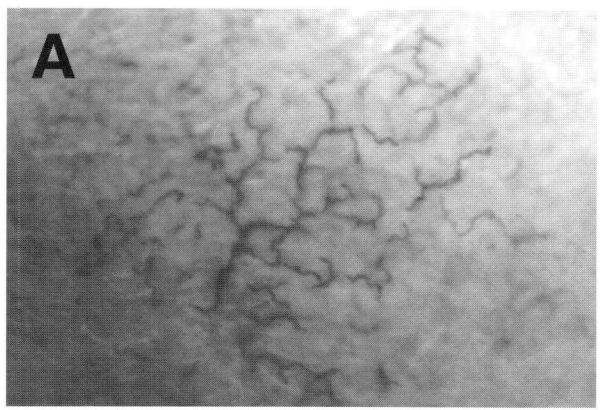 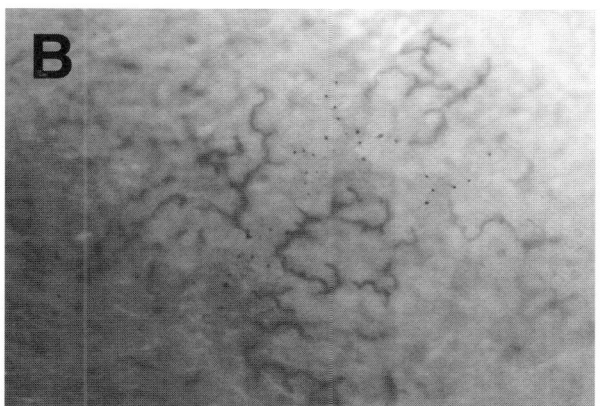

Figure 66
A) Linear telangiectases on the cheek of a 47-year-old woman, before treatment.
B) About 10 days after treatment, tiny crusts are seen. Typically, crusts are the size of a pin point. Notice that treated vessels are no longer visible; this initial treatment was successful. Subsequent treatments removed the remaining telangiectases.

Allow sufficient time for crusts to form between treatments. Medicated salves or powders are unnecessary following treatment, and should only be used if infection is seen. Furthermore, the skin must be calm at each treatment, not red or swollen from the last treatment. Be sure all crusts have sloughed off before treating the same area again (Figures 66 and 67).

AFTER-CARE ADVICE

After-treatment care is critical for success. Nothing must be done to open the coagulated blood vessel segments. Anything that increases the skin's blood flow or stretches the skin must be avoided for at least twenty-four hours.

Patients must not stretch their skin to observe the treated area. They must not sunbathe, exercise, take a sauna, wash their face or apply make-up for at least twenty-four hours following treatment.

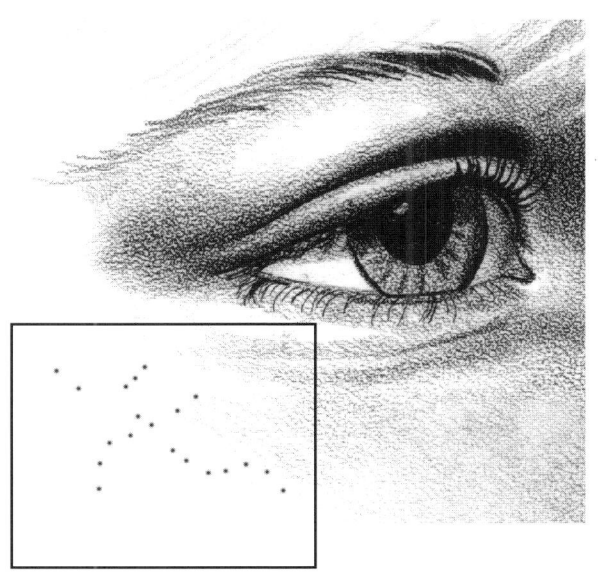

Figure 67
Crusts should be about 0.5 millimeter in diameter. Do not treat the same area again until crusts have fallen off, inflammation has completely resolved and the skin appears normal.

POST-INFLAMMATORY HYPERPIGMENTATION

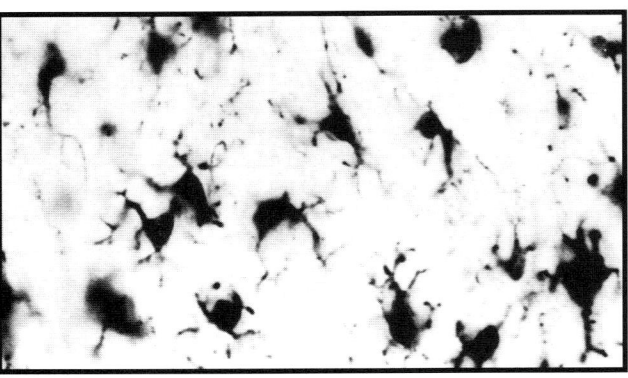

Figure 68
Melanocytes (pigment producing cells) as seen on the underside of the epidermis. Inflammation can stimulate these cells to produce freckle-like marks. (Photo is courtesy of William Montagna, Ph.D.)

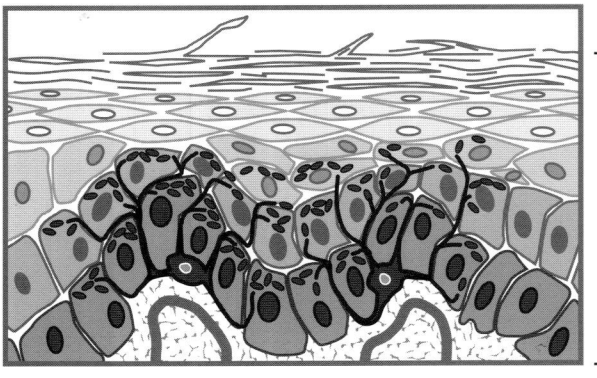

Figure 69
When stimulated by sun or inflammation, melanocytes make melanosomes (pigment bundles). Epidermal cells then phagocytize (ingest) the melanosomes and produce a suntan or PIH.

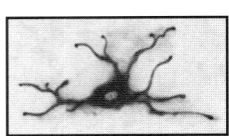
MELANOCYTE

Melanocytes are unique cells located in the basal layer of the epidermis that, when stimulated by sunlight, produce melanin—skin pigment (Figures 68 and 69). If treatment with the blend or other high frequency device inflames the basal layer of the epidermis, the melanocytes can be stimulated to produce pigment. This phenomenon is defined as "post-inflammatory hyperpigmentation," or PIH. Because only a small spot of skin is stimulated when removing telangiectases, the hyperpigmentation often resembles a tiny dark spot or a freckle (coloration ranges from yellow to dark brown or black).

Because telangiectases are just under the surface of the skin, the epidermis and basal layer are always affected by the treatment. However, hyperpigmentation is almost never a problem with fair-skinned people: their melanocytes are not significantly stimulated by inflammation. Since the majority of patients with telangiectases are fair-skinned, the problem of PIH is rarely experienced.

Occasionally, a patient with a dark complexion or olive skin will be treated. In such cases, PIH is probable. Be sure to warn the patient to expect this problem. Luckily, post-inflammatory hyperpigmentation is temporary. Although the marks can last up to six months or longer, they eventually fade. (See photo 8, page 66.)

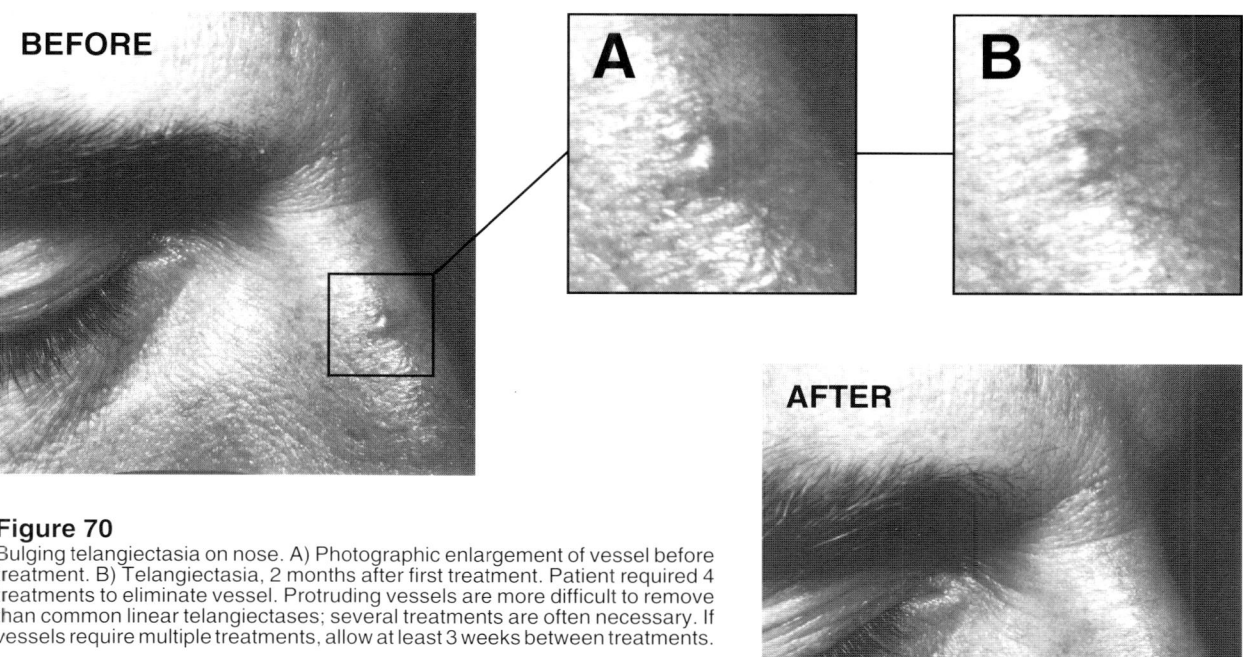

Figure 70
Bulging telangiectasia on nose. A) Photographic enlargement of vessel before treatment. B) Telangiectasia, 2 months after first treatment. Patient required 4 treatments to eliminate vessel. Protruding vessels are more difficult to remove than common linear telangiectases; several treatments are often necessary. If vessels require multiple treatments, allow at least 3 weeks between treatments.

PROPER LIGHTING

Proper lighting is important when treating telangiectases. An incandescent lamp (hot light), such as the common light bulb, makes seeing facial capillaries difficult because it produces surface glare on the skin. A dental lamp or halogen lamp also produces glare. If the skin is oily, surface glare usually becomes even worse.

If surface glare is a problem, buy a pair of "clip-on" polarized sunglasses that fit over your regular glasses; try to find the very lightest shade lenses. You will find that polarized lenses reduce surface glare and make telangiectases more visible.

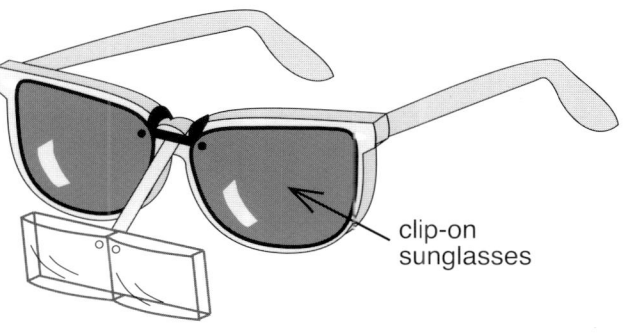

A fluorescent lamp (cool light) is recommended for telangiectasia work, because this lamp creates little surface glare. Purchase a small, inexpensive fluorescent desk-type lamp for this work. Polarized sunglasses can be used with the fluorescent lamp to nearly eliminate surface glare.

ADVICE TO AVOID TELANGIECTASIA

Some sensible advice can be given to reduce the recurrence of telangiectases. The patient should avoid prolonged sun exposure and wear a sunscreen lotion when needed. Since temperature extremes might hasten the development of telangiectases, temperature extremes should be avoided; especially long exposure to the cold. The patient should refrain from careless squeezing of blackheads. White, thin-skinned people must be especially careful because they are prone to develop telangiectasia.

Certain experts believe that patients should avoid excessive heat to the face and learn the recommended way of taking a sauna. They say a warm shower should be taken prior to entering a sauna to get the skin used to the heat. The face should be covered with a damp towel for the first few minutes while the skin becomes adjusted to the heat.

Several medical authorities suggest that vitamin C strengthens blood vessels and thus reduces the risk of telangiectases. Although vitamin C is not a prescription drug, only a medical doctor can "prescribe" vitamin C. However, the therapist might recommend that the patient take an "adequate amount" of the vitamin, or refer the patient to one of the many books that deal with vitamin C.[21]

---

21) Popular books on Vitamin C:
Linus Pauling: *Vitamin C and the Common Cold* (1970).
Emanuel Cheraskin: *The Vitamin C Connection* (1983).
Irwin Stone: *The Healing Factor: Vitamin C* (1972).

TREATMENT SUCCESS

As with all procedures, some patients will not respond to treatment—and, for no apparent reason. More than 90 percent, however, will find the treatments successful.

If a patient is prone to telangiectasia formation because of heredity, treatments will only be temporary. New dilated capillaries will eventually form; the patient must be advised of this possibility. However, it could be many months or even years before new telangiectases develop. A patient with thick strong skin who got dilated vessels from an injury will respond permanently to treatment.

Be advised that patients often become too fastidious. They may demand the removal of tiny capillaries that are difficult to see. Advise the patient that only conspicuous marks should be removed, and that to do more is a waste of money.

CHARGING FOR TREATMENTS

If only a few capillaries are to be treated, a one-time fee (not based on treatment time) can be charged. For example, a set charge can be made for a simple lesion requiring only a few minutes of treatment time. With extensive work, such as an entire upper chest, charging by treatment time seems appropriate.

Removing telangiectases is gratifying to both the therapist and the patient, because results are seen immediately. You will find that treating telangiectases is a welcome psychological lift for both the patient and therapist. As a result, offering telangiectasia treatments can enhance and stimulate your entire practice.

QUICK REFERENCE CARD

Photocopy or cut out this card and keep it in your work area. Use it as a quick reference until you become completely familiar with all the procedures.

NEEDLE SIZE:	.002 for tiny thread size vessels. .003 and .004 for medium to large vessels.
MACHINE POWER:	HF: Low power for thread size. Medium power for larger vessels. DC: 0.2 to 0.3 milliamperes for thread size. 0.3 to 0.4 milliamperes for larger vessels.
• *BASIC PROCEDURE:*	Insert with DC only. Coagulate with HF and DC. Remove needle with DC only.
• *MODIFIED BASIC PROCEDURE:*	Insert with DC Coagulate with HF only. Remove needle with DC only.
EXTENSIVE CONDITION:	Insert with HF and DC. Coagulate with HF and DC. Remove needle with HF and DC.
SPIDER TELANGIECTASIA:	Use **basic procedure** unless bubbles seen.
DOT-TELANGIECTASIA:	Use **basic procedure** unless bubbles seen. Coagulate as spider center.
BLOOD-MOLE:	Use **basic procedure** unless bubbles seen.
RED BLOTCH:	Treat randomly.
DC ONLY PROCEDURE:	Try for extremely small vessels only. DC : 0.2 to 0.3 milliamperes.

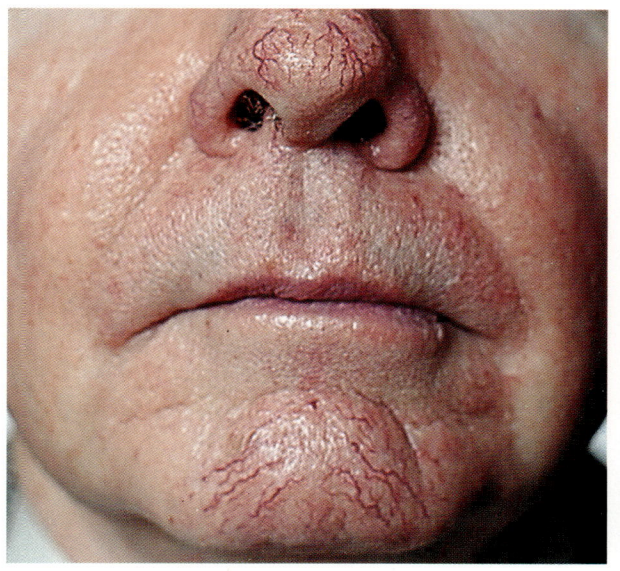

Photo 1
Patient: 56 year old man (linear telangiectasia).
Medical diagnosis: hereditary telangiectasia.
Patient had been treated with the hyfrecator: notice scar tissue on center area of the nose

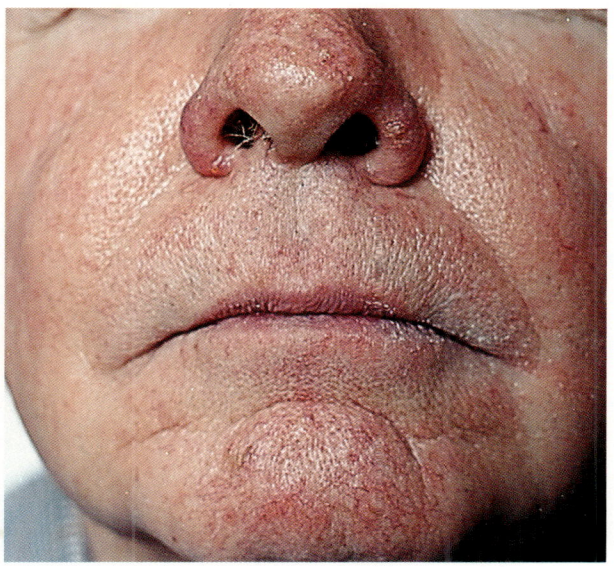

Photo 2
Six treatments of 5 minutes each were required to eliminate vessels on the chin and nose. Scar tissue from hyfrecator is still visible.

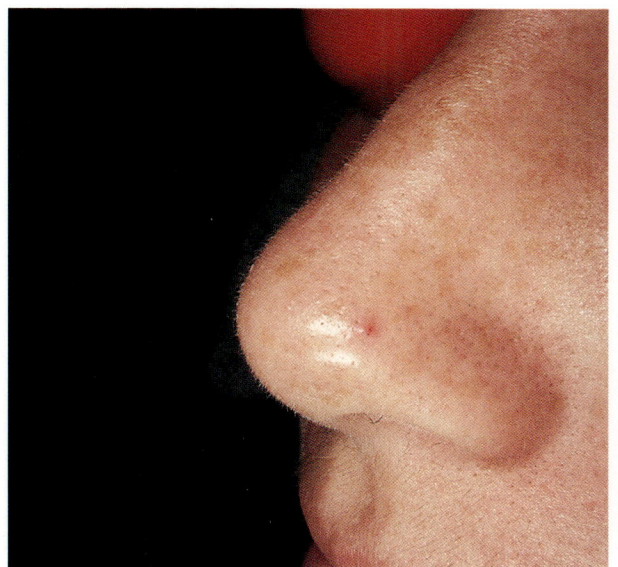

Photo 3
Patient: 16 year old girl (dot-telangiectasia).
Medical diagnosis: telangiectasia of unknown cause, possibly injury induced.

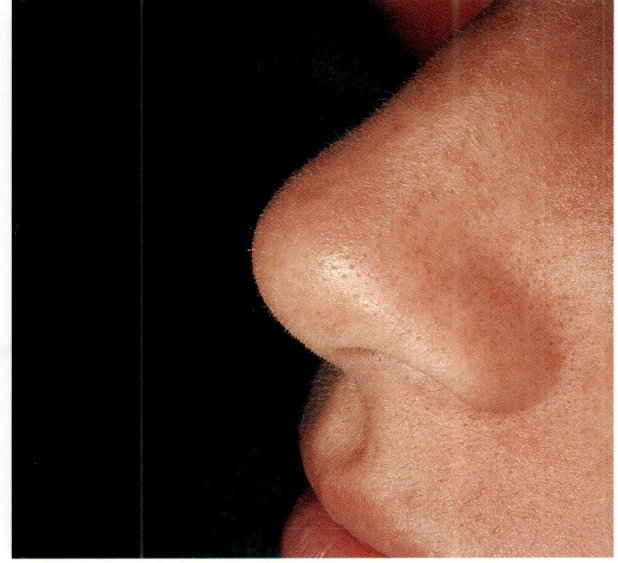

Photo 4
Only one treatment was required. Although patient had very moist and delicate skin, no scar tissue is visible following treatment with the Blend.

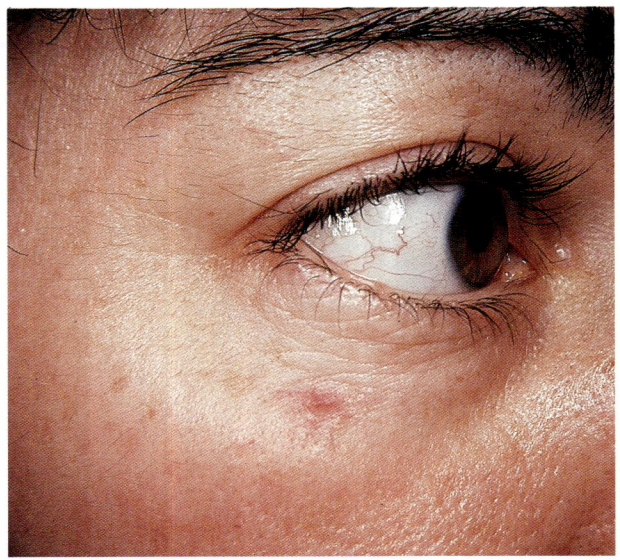

Photo 5
Patient: 26 year old woman.
Medical diagnosis: hereditary telangiectasia.
Patient before treatment.

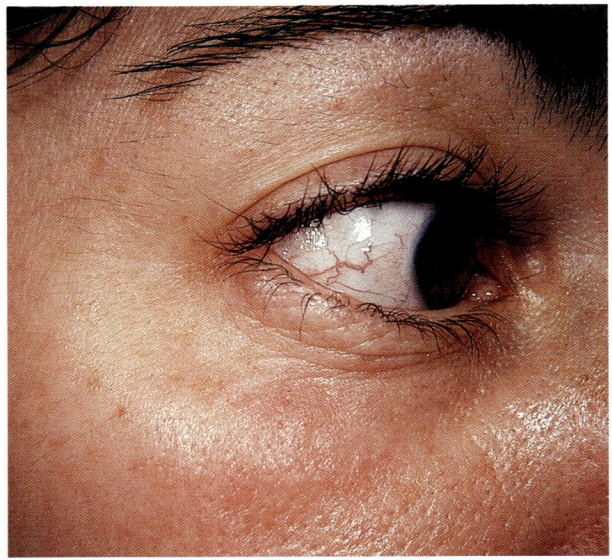

Photo 6
Patient immediately after treatment. In most cases vessel is not visible after treatment. Very little edema or erythema is seen. Tiny crusts form in 24 to 48 hours following treatment.

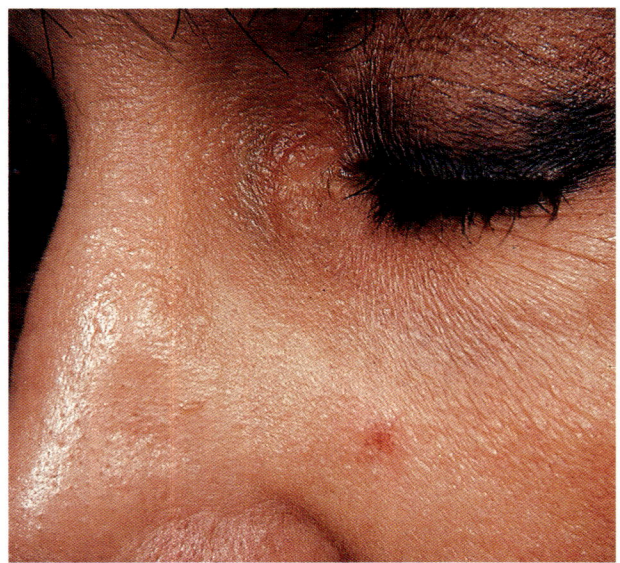

Photo 7
Patient: 42 year old woman.
Medical diagnosis: telangiectasia of unknown cause. Because patient is of Greek heritage and has olive skin, hyperpigmentation was expected.

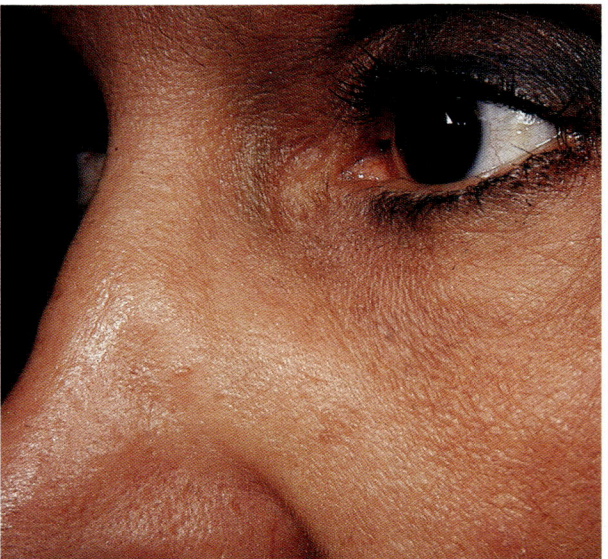

Photo 8
Patient 8 weeks after treatment. Vessel required only one treatment but hyperpigmentation did take place (note tiny freckle). Mark was not visible 4 months later.

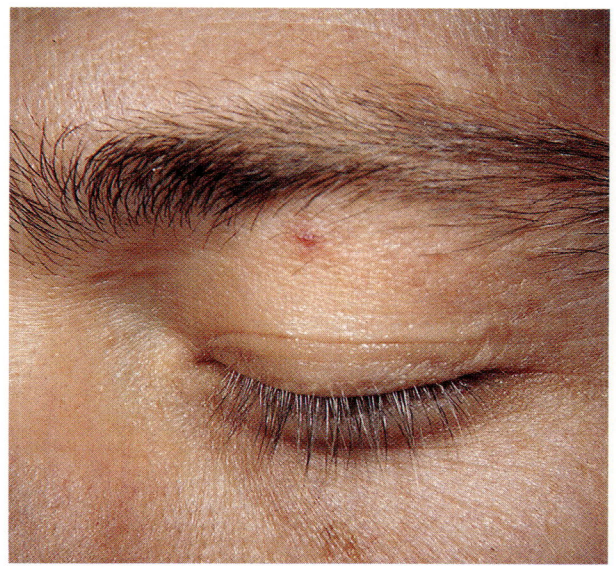

Photo 9
Patient: 30 year old man.
Medical diagnosis: congenital telangiectasia.
Lesion was rather large and protruded above skin line. Patient claimed to have had lesion since birth.

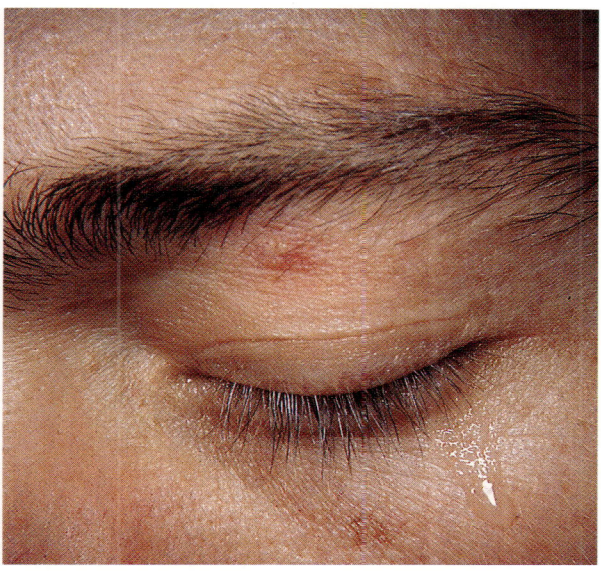

Photo 10
Patient immediately after treatment. (I wonder if those are "tears of joy?")

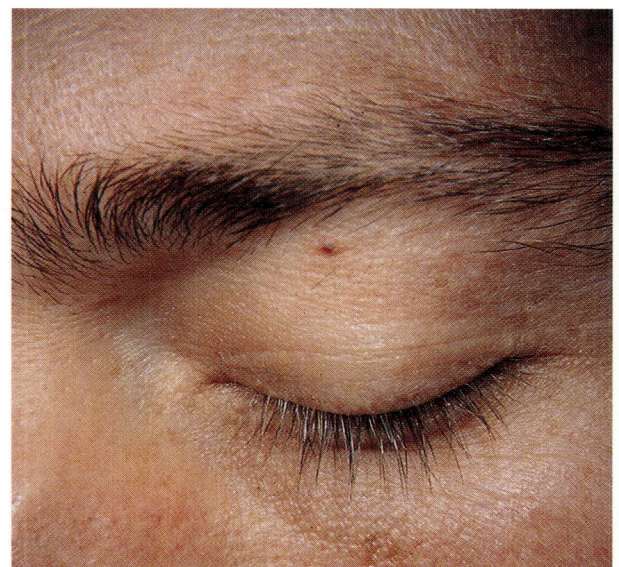

Photo 11
Patient 24 hours later: vessel "leaked" blood. Treatment, however, was successful---only one treatment was necessary.

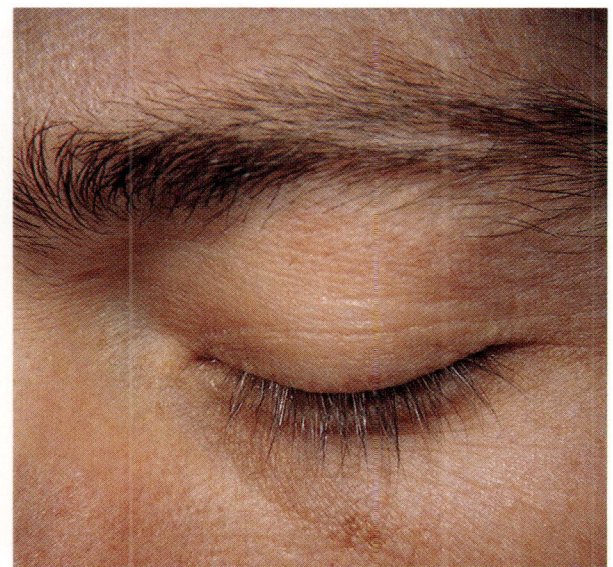

Photo 12
Patient 6 months after treatment. Vessel was permanently eliminated, scar tissue is not seen.

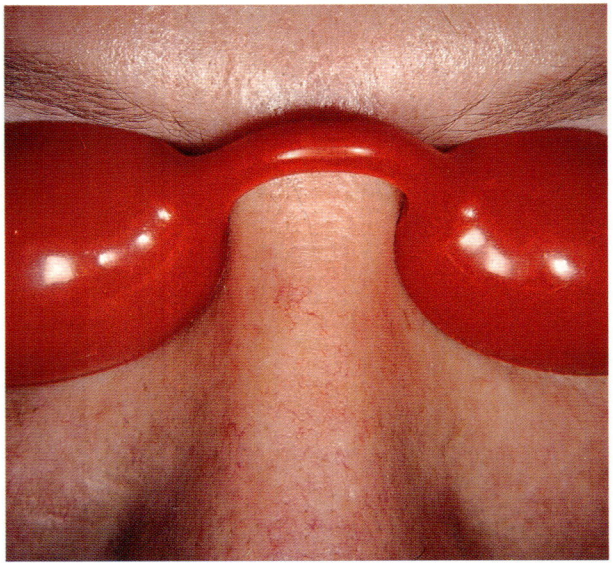

Photo 13
Patient: 35 year old woman of Scottish heritage.
Medical diagnosis: hereditary telangiectasia.
Photo shows the most common form of hereditary telangiectasia seen on people with thin white skin.

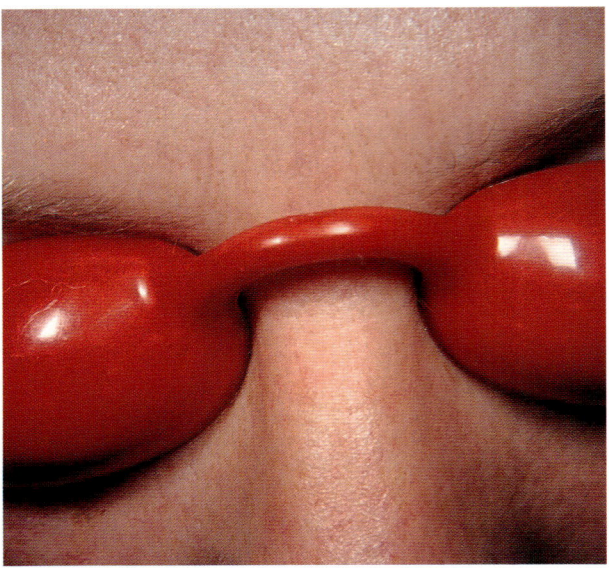

Photo 14
Area is much improved following treatment. Treatment of this type of telangiectasia is nearly 100% successful in all cases.

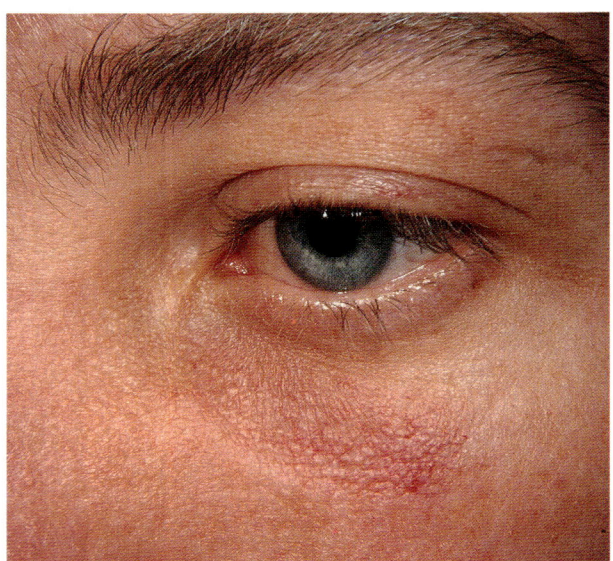

Photo 15
Patient: 24 year old man.
Medical diagnosis: congenital telangiectasia (birth mark). Patient claimed that many family members have similar vascular birthmarks under the eyes.

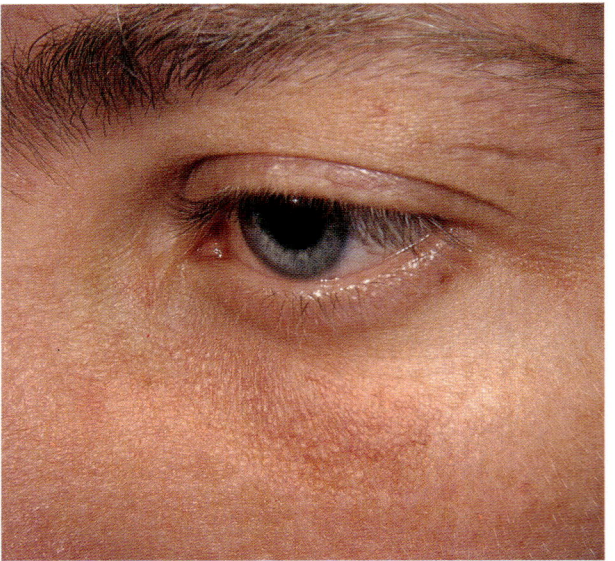

Photo 16
Patient after 3 treatments. Notice lesion is about half gone. Mark was eliminated after three more treatments.

GLOSSARY

GLOSSARY

AGE SPOTS
Also called "liver spots," or solar lentigines, these are yellow, brown or black spots of pigment in the skin. These marks typically develop with age and frequently appear on the hands, arms and face.

AIDS
Acquired Immunodeficiency Syndrome: caused by the HIV virus. A disease that destroys the immune system and causes many opportunistic infections such as pneumocystis carnii pneumonia, CMV, candidiasis and malignancies such as Kaposi's sarcoma. Disease is pandemic and transmitted by body fluids such as blood and semen.

ANESTHETIC
A generic term for agents that temporarily depress nerve function causing the loss of pain and sensation. There are many different anesthetics available. They are used in all forms of medical treatment as well as in dentistry.

ANGIOMA
A swelling or tumor caused by blood vessel dilation (as in hemangioma) or lymphatics (as in Lymphangioma).

ARGON
An inert gas (element) found in the earth's atmosphere. Atomic number is 18, atomic weight is 39.95.

GLOSSARY

ARTERIOLE
(Also arteriola) A very small muscled artery that connects with the capillary network.

ASPIRIN
Common pain relieving drug used throughout the world. Acetylsalicylic Acid $C_6H_4(OCOCH_3)-COOH$. Drug is analgesic, anti-inflammatory and antipyretic. Can cause stomach disorders and acts as a blood anti-coagulant.

BASAL CELL CARCINOMA
Skin cancer arising from epithelial (skin) cells. Condition is usually not fatal and rarely metastasizes (unlike melanoma). Is believed to be caused by excessive and prolonged sun damage to the skin.

BASAL LAYER
Active layer of the epidermis containing "growth cells" and pigment. Basal layer also forms hair follicles.

BIRTHMARK
General term denoting a minor skin blemish from the time of birth. Moles, patches of pigment or blood vessel marks are common forms of birthmarks.

CAPILLARY
The smallest of blood vessels. Thread-like vessels that "feed" the cells of the body with nutrients and oxygen and carry off waste products.

GLOSSARY

CARCINOMA
(Greek: *karkinos,* cancer)
Any of the various types of malignancies of epithelial tissues. Most frequently in the skin and intestine, stomach and prostate gland, and breast and cervix in women.

CATAPHORESIS
A process that uses the positive pole of galvanic (DC) current to introduce positively ionized substances into the skin. Soothes the skin, reduces redness (is a vasoconstrictor), firms tissues and is a mild germicidal (acid reaction on the skin is said to kill germs). Generally applied with a "cataphoresis roller" implement.

CAUSTIC SOLUTION
A liquid chemical causing a burn.

CAUTERIZE
To intentionally burn the skin with a device or substance. Causes scarring, burning or cutting of the skin by heat, electrical currents or chemicals.

CELL
The basic living structure of all plants and animals. Composed of protoplasm (a jelly-like substance), a nucleus (containing genetic material) and enclosed in a fine membrane.

GLOSSARY

CHEMICAL PEEL
Used in cosmetic surgery. Procedure is a form of "dermabrasion." Caustic material is applied to the face or other body part to burn off layers of the skin. Regenerating skin tissue usually has fewer wrinkles and fewer age spots. There are many different types of chemical peels. Usually administered by a medical doctor.

CHRONIC
(Greek: *chronos,* time)
Term denotes any disease or condition of slow progress and long duration. Many chronic conditions or diseases are not curable. They may or may not be fatal.

CIRCULATORY SYSTEM
The entire blood system including all vessels and organs that move blood throughout the body in a circular course.

CLOT
A coagulation. A soft pliable mass formed when blood or lymph gel.

COAGULATION
Blood forming a solid mass. A "clotting."

COMEDO (pl. COMEDONES)
A blackhead. A plug of sebaceous oil and skin cells filling the sebaceous gland opening. Usually has a blackened top caused by accumulated skin cells and rancid sebaceous oil.

GLOSSARY

CONGENITAL
(Latin: *congenitus,* born with)
A condition existing at birth that may be hereditary or caused by a trauma during pregnancy.

CROCUS CLOTH
A very fine abrasive cloth used for polishing. Sometimes called a "jeweler's cloth."

DERMATOMYOSITIS
A chronic and progressive disease. Muscular weakness and skin rashes are common hallmarks. Telangiectasia of the face and edema (swelling) of eyelids takes place.

DERMIS
The "true" skin. Lying below the epidermis, the dermis contains blood vessels, connective tissues and nerves.

ELECTROCOAGULATION
Coagulation of tissue or blood, by means of a high frequency apparatus such as an electrocautery, hyfrecator of HF generating electrolysis device.

EMBOLISM
(Greek: *embolisma,* a piece)
An obstruction of a blood vessel by a clot, bacteria or foreign particles.

GLOSSARY

EPIDERMIS
The outer layer of skin composed of flat dead cells and active epithelial cells. The body's primary shield against infection from bacteria, viruses, funguses, organisms or materials.

EPINEPHRINE
Adrenaline. A neurohormone that stimulates the nervous system. Causes increased heart rate, increased pain perception. Used in local anesthetic preparations as a vasoconstrictor to attain longer periods of localized anesthesia. Also used as a drug for many conditions including the treatment of asthma.

ESTROGEN
"Female hormone." May be natural or synthetically produced. Hormone is produced by the ovaries, testes, placenta, adrenal cortex and some plants. Hormone causes secondary female characteristics. Is used medically to aid many conditions and diseases in both men and women.

FACE LIFT
Surgical procedure by either a plastic surgeon or a cosmetic surgeon. Excess facial tissue is removed surgically from the face. Skin is pulled back creating a more youthful appearance.

GLOSSARY

FLUORESCENT LAMP
Invented in 1895-1900. A tubular lamp giving light by electrical stimulation of a coating of phosphors on the inside of the tube. Creates daylight to blue light.

HALOGEN LAMP
A gas filled high-intensity lamp. The filament is made of tungsten and becomes very hot, vaporizing minute amounts of iodine in the lamp. This creates a very intense light of natural daylight color.

HEREDITY/HEREDITARY
(Latin: *hereditas,* inheritance)
Characteristics passed on from parent to child. Determined by "inherited" genes.

HIGH BLOOD PRESSURE (HYPERTENSION)
Caused by several diseases, life style (smoking), and hardening of the arteries. Chronic condition sometimes leads to heart attack. Treatable with diet, exercise and medication.

GLOSSARY

HIGH FREQUENCY
An oscillating electrical current which ranges from 3 to 40 megahertz. Also called: radio wave, diathermy, short wave (SW), radio frequency (RF) and thermolysis. High frequency heats human tissue by friction: caused when negative charged electrons of the tissue are "magnetically" vibrated. High frequency can warm, coagulate or desiccate tissue. Main current used in skin surgery such as mole removal, hair removal and capillary removal.

HORMONE
(Greek: *hormao*, to rouse, set in motion)
A natural substance produced in various glands in the body. Substances are carried by the blood and affect other glands, functions, structures and systems of the body. The hormone estrogen can cause telangiectasia.

HYDROGEN
(Greek: *hydro-gen*, water producing)
An abundant combustible gas found in the earth's atmosphere; with oxygen it forms water. The lightest element. Symbol H, atomic number 1, atomic weight 1.0079.

GLOSSARY

HYPERPIGMENTATION
Excess pigment in the skin or other tissue. A problem with dark or olive complexions because treatment with any coagulation device often causes the condition. Lesions appear as spots or dark freckles. Hyperpigmentation can take between 6 months to one year to fade.

INCANDESCENT LAMP
The most commonly used lamp, a "light bulb." Light is emitted by a heated filament, usually in an evacuated or gas filled bulb. Bulb creates significant heat.

INCISION
A surgical cut made with a surgical knife such as a scalpel, laser or high frequency cutting device.

KELOID
A thick and nodular mass of scar tissue. Consisting of collagenous fibrous tissue, these lesions can follow injury, burns or diseases such as acne. The tendency to form keloids is hereditary.

LASER *(Light Amplification by Stimulated Emission of Radiation).*
A general term defining a device that concentrates light into a narrow parallel beam. There are many types of lasers used for many diverse purposes, including: surgery, photography, entertainment and warfare.

GLOSSARY

LESION (Latin: *laesus,* to injure)
A very basic term used in medicine. A structural or functional change in tissue caused by disease or injury. Lesions may be primary or secondary. Almost any change on the skin can be called a lesion (such as a burn, abrasion, cut, dilated vessel, age spot, etc.)

LUMEN (Latin: *lumen,* window)
The open space in a tubular structure. Intestinal lumen. The lumen of a vein.

LUPUS ERYTHEMATOSUS
(Latin: *lupus,* wolf)
The term Lupus defines various diseases. The disease, in some forms, causes the skin to appear as if "gnawed." Lupus Erythematosus particularly affects skin, especially of the face. Erythema, hyperkeratosis, plugged follicles and telangiectasia are commonly displayed.

MELANIN
(Greek: *melas,* black)
Brown to black pigments produced by specific cells of the body. These pigments are found in hair and skin.

GLOSSARY

MELANOCYTES
Cells in the skin responsible for producing melanin: skin pigment. Cells are located at the junction of the epidermis and dermis. The cells use a branching process to distribute pigment to the epidermis. Cells are primarily stimulated by sunlight but may be stimulated by inflammation in certain individuals.

MILLIMETER
A unit of metric measurement equal to one thousandth of a meter or 0.03937 inches. About 25 millimeters make one inch. The metric system of measurement is the most standard system of measurement (except in the United States). It is used exclusively in the sciences.

MOLE
(Latin: *macula,* a spot)
Scientific term is *Nevus Pigmentosus.* A congenital mass of pigmented skin cells. Moles can take many different forms. Some types should be medically checked because they can become skin cancer.

NAEVI (OR NEVI)
Plural of nevus: a mole.

NECROSIS
(Greek: *nekrosis,* death)
Cell death or death of part of an organ or tissue resulting from a specific injury.

GLOSSARY

ORAL CONTRACEPTIVE
A drug or preparation taken by mouth that prevents conception and pregnancy. These drugs are used by women and often contain hormones which can cause telangiectasia in some individuals.

POLARIZED LENSES
A specially designed lens that allows light from only one direction to pass through. Other angles of light are reflected. These lenses are used to reduce surface glare. They are popular sunglasses for driving automobiles or for water sports. Any activity where surface glare can be a problem can be aided by wearing polarized lenses.

PORT WINE BIRTHMARK
A hereditary blood vessel birthmark that is highly vascularized. Color ranges from dark brown to bright red (the color of red port wine). The most famous port wine birthmark proudly sits atop the head of Mikhail Gorbachev, former president of the Soviet Union.

RAYNAUD'S PHENOMENA
Described in 1880 by Maurice Raynaud of France, this is a vascular disorder of unknown cause. The disease can cause numbness of the fingers, toes, ears and nose. Areas of the skin can blanch. Telangiectasia can be seen. Episodes are triggered by stress or exposure to cold.

GLOSSARY

RETINA
The innermost coating of the eyeball. A network of nerves that receive visual light. The retina can detach from the eyeball, causing blindness. Laser is used to "weld" the detaching retina back in place. Laser surgery for detached retina is extremely successful.

ROSACEA
(Latin: *rosaceus,* rosy)
Vascular and follicular dilation of the nose and cheeks. Acne rosacea involves persistent erythema of sebaceous (oil) glands of the face with severe pustules and telangiectases.

SAUNA
A wet or dry heat bath that induces perspiration. There are many different types of saunas. Popular in cold climates.

SCAB
A crust of coagulated blood and serum (sometimes pus) that has collected on the surface of a wound. Scabs form on breaks in the skin or following burns.

SCAR
Fibrous tissue growth that replaces normal tissue destroyed by injury or disease.

GLOSSARY

SCLERODERMA
A chronic progressive disease marked by thickening and swelling of the skin with fibrous tissue growths. Epidermis eventually atrophies. Disease often shows telangiectasia of the face and body.

SCLEROSIS
(Greek: *sklerosis,* hardness)
A hardening of tissues or cells. An increase in connective tissues replacing more normal tissue.

SCLEROTHERAPY
Treatment for varicose veins in which sclerosis is created in blood vessels to divert blood flow by injecting a "hardening" solution (solution is called sclerosant).

SODIUM HYDROXIDE (NaOH)
Also called caustic soda or lye. A base, sodium hydroxide can dissolve human tissue. Sodium hydroxide is produced at the negative pole (needle) when it is introduced into human tissue.

STELLATE
(Latin: *stella,* a star).
Star-shaped, as in a telangiectasia that may be shaped like a spider or a star.

THROMBOSIS
(Greek: *thrombos,* a clot)
A clotting of blood within the blood vessel.

GLOSSARY

VARICOSE VEIN
A vein that has become severely dilated or swollen. These veins are found mainly on the legs. More common in women.

VEIN
General term for blood vessels that return blood to the heart and lungs.

VENULE
A minute vein connecting a capillary and vein. These vessels are continuous with capillaries.

VITAMIN C
(Latin: *vita,* life)
An organic substance present in small amounts in natural foodstuffs, especially fruits and vegetables, that are essential to human life. May also be synthetically produced. Vitamin C is ascorbic acid.

X-RAY
(Also called Roentgen ray) Electromagnetic radiation produced in a vacuum tube containing a target anode and heated cathode. X-ray has many uses in medicine and science.

INDEX

INDEX

A
"Ace" bandage: 29
acetylcholine: 22
adrenaline: 22
after-care: 57
AIDS: 20
alcoholism theory: 22
angiogenesis: 24
angle of insertion (needle): 42
ankles: 29
argon laser: 32
arteries: 26
arterioles: 26, 44
aspirin: 54

B
basal cell carcinoma: 19
basal layer: 58
basic procedure (blend): 38-39
bent needle: 42
blackhead: see comedo
blood mole angioma: 48
blood spots: 47
blood vessel birthmark: 32
blotchy areas: 49
bra: 48
bruising: 30
bubbles (blend procedure): 41
butcher's cheeks: 23

C
capillaries: 26
carborundum: 36
cataphoresis: 56
caustic solution: 30
charging for treatments: 61
chemical peel: 18
chronic hepatitis: 20
clip-on sunglasses: 59
clot: 24, 34, 40
comedo (blackhead): 17, 60
compression bandage: 29
concentrated insertions: 55
crusts: 29, 31, 32, 56-57

D
dark skin: 58
DC: 39
DC alone (technique): 50
dermatomyositis: 20
direct current: 39
diseases: 19-20
dotted line: 29
dot-telangiectasia: 47

E
edema: 55
EMLA: 53
epinephrine: 53
epidermis: 58
exercise: 57
extensive condition: 43

INDEX

F
face lift: 19
face-technic: 15, 29, 37, 38
fluorescent lamp: 59

H
heredity: 23
HF: 31
HF epilator: 33
high blood pressure: 19
high frequency: 31
Hokkaido: 23
hormone: 18
hyfrecator: 31
hyperpigmentation: 58

I
ice: 54
increased blood flow: 52
inherited telangiectasia: 23

K
keloid: 19

L
laser: 32
leaking vessel: 46
lighting: 59
linear vessels: 40
lips: 47
local anesthetic: 53
low angle of insertion (needle): 42
lupus erythematosus: 20
lye: 39

M
make-up: 57
melanin: 58
melanocytes: 58
milliamperes: 37
modified basic procedure: 41
mole: 19

N
needle size: 36
needle sticking: 34
nevus (mole): 48
nose: 47

O
olive skin: 58
oral contraceptives: 18
overheated skin: 55
overtreatment: 42-43

INDEX

P
Peru: 23
PIH: 58
polarized glasses: 59
port wine birthmark: 33
post-inflammatory hyperpigmentation: 58
pregnancy: 18

R
Raynaud's phenomenon: 20
recanalization: 24
red blotchy areas: 49
Retin-A: 18
retina: 32
rosacea: 20

S
sauna: 57, 60
scars: 32, 34, 40
scar tissue: 19
Schnappsnase: 23
scleroderma: 20
sclerosant: 30
sclerosis: 30
sclerotherapy: 28, 30
sodium hydroxide: 39
spider angioma: 44
spider telangiectasia: 44
spider veins (legs): 28
stretching the skin: 52
sunbathe: 57
sun exposure: 17
sunglasses: 59
sunscreen lotion: 60
surgery: 28
swelling: 55

T
tapping technique: 34
telangiectasia (definition): 16
temperature theory: 21-22, 60
thrombosis: 30
tortuous vessels: 40
treatment time: 54

UVWXYZ
varicose veins: 28
veins: 26
venules: 26
vitamin C: 60
wash face: 57
Xylocaine: 53
yellow dye laser: 32

THE BLEND METHOD

THE BLEND METHOD

The so-called blend method is a specific modality used in electrolysis for permanent hair removal. The method unites high frequency and direct current in a conductive needle to coagulate the hair follicle. The blend method was invented by Henri St. Pierre and Arthur Hinkel in 1948. Hinkel, however, developed the technique into an organized system. (See *Real World Electrology: The Blend Method*, 1995, by Michael Bono, for a full explanation of the blend method of electrolysis.)

Today, many different blend machines are made worldwide: Canada, Japan, the United States and Europe. Unfortunately, not all of these epilators can be used for telangiectasia treatment. Before purchasing an epilator for this procedure, consider certain criteria.

1   The currents of the epilator must not be automatically controlled! Many machines feature "computer controlled" currents—these must not be used! Both the HF and the DC currents must be totally independent and manually controlled. Both currents must be controlled by an individual footswitch (one for each current). Note: some computerized machines can be manually controlled—ask the manufacturer and try out the epilator!

THE BLEND METHOD

2   HF power output is critical! Some blend epilators produce too much current. For telangiectasia work, the HF current must have a range of 35 to 55 volts (p-p) under normal human body resistance. Be sure to ask the manufacturer if these current levels are attainable. (Hint: at zero on the HF dial, the machine should produce no perceptible current. If the machine is already producing current at zero, it is probably too high and cannot be used.)

3   DC current is seldom a problem; most machines can produce between 0.2 and 0.4 milliamperes, the amount of current needed for telangiectasia removal. However, some machines produce a "faradic effect" in the DC current which can produce a mild shock. Test the machine for tiny shocks in the DC circuit (see pages 340-341 in *Real World Electrology: The Blend Method*, Bono, 1995).

If you are considering purchasing a new blend epilator, be sure to ask the manufacturer if the machine has proper output levels. Try out the epilator before you buy! It is not the scope of this textbook to include the blend technique. Refer to current operational manuals, textbooks and state licensed or accredited electrology schools for more information.

TEST YOUR KNOWLEDGE

Multiple Choice Questions

1. The port-wine birthmark is best treated by: a) the blend, b) laser, c) hyfrecator, d) saline injection.

2. An arteriole is: a) a small vein, b) a small artery, c) the same as a capillary, d) any vessel carrying oxygen deficient blood.

3. Most cases of telangiectasia are: a) caused by disease, b) the result of high blood pressure, c) hereditary, d) caused by frequent exposure to the cold.

4. Patients with injury caused telangiectases: a) usually respond well to treatment, b) are careless with their skin, c) generally heal poorly, d) do not respond well to treatment.

5. Telangiectasia seen in pregnancy: a) may be safely treated, b) indicates a weakness in the blood vessels, c) usually resolves 6 weeks after delivery, d) must be diagnosed by a doctor.

6) Telangiectases in scar tissue: a) may be treated with medical approval, b) indicates failure of the skin to heal properly, c) is never seen following face lift surgery, d) may not be treated.

7. Every patient with telangiectases: a) must sign a release statement before treatment, b) must be seen by a doctor to rule out disease, c) is often fastidious, d) has a hereditary predisposition for forming dilated capillaries.

8. To remain current on telangiectasia treatment, you should: a) practice frequently, b) read related literature that deals with telangiectasia treatment, c) subscribe to medical journals that deal with vascular surgery, d) none of the above.

9. Dilated facial capillaries appear more frequently in: a) people with liver disorders, b) people who eat hot and spicy foods, c) people who consume alcohol, d) white thin-skinned Northern Europeans.

10. Alcoholism can cause telangiectases because: a) years of drinking keeps blood vessels dilated, b) the liver can become damaged and consequently interfere with normal hormone levels, c) heat to the face from drinking enlarges the vessels.

11. Most medical authorities consider telangiectases: a) a significant medical problem, b) disease related, c) a congenital weakness in the blood vessel walls, d) a nuisance to treat.

12. Eliminating telangiectases from a patient's face: a) is of minor importance to the patient's health, b) eliminates the need for make-up, c) indicates the patient is fussy, d) can greatly help self-confidence because society thinks dilated blood vessels are caused by alcoholism.

13. All treatments to eliminate telangiectases: a) cause tissue contraction, b) attempt to clot the vessel, c) must be concerned with the direction of blood flow in the capillary.

14. After scar tissue forms in the treated vessel: a) blood may recanalize through the scar tissue, b) blood will never again flow through the capillary, c) tiny dot-like scars will be seen.

15. A vein: a) brings oxygenated blood to the cells, b) connects arteries to capillaries, c) returns blood to the heart and lungs, d) cannot be treated by any method available.

16. Using the blend method, you should: a) work exclusively on the legs. b) limit your work to capillaries only, c) treat varicose veins, d) limit your work to tiny blood vessels on the face, neck and upper chest.

17. The commonly used hyfrecator sometimes causes visible scars, because: a) the needle used is too thick, b) doctors habitually use too much HF current, c) the HF of the hyfrecator cannot be adjusted low enough.

18. The key difficulty in using an HF only epilator is: a) the needle often sticks and pulls out the coagulated segment, thus reopening the vessel, b) the HF current is too drying, c) few therapists have access to this type of device.

19. When treating telangiectases with the blend method, remember that HF coagulation: a) takes place along the entire length of the needle, b) take at lease 10 seconds to form, c) cannot be predicted, d) is concentrated at the tip of the needle.

20. For most telangiectasia work, the DC is set at: a) 1 milliampere, b) 2 to 4 milliamperes, c) 0.2 to 0.4 milliamperes.

21. The **basic procedure** of the blend method: a) uses high HF and low DC, b) uses high HF and high DC, c) uses low HF and low DC, d) uses low HF and high DC.

22. Using the **basic procedure** blend technique: a) insertion is made with both currents off so the vessel will not bleed, b) insertion is made with HF on, to quickly coagulate the epidermis, c) insertion is made with DC on, so that the needle easily penetrates into the vessel lumen.

23. When treating linear vessels, try to coagulate the entire length and a) place insertions next to each other, b) place insertions at least 2 to 3 millimeters apart, c) place insertions at the beginning and end of the vessel.

24. If bubbles appear in a vessel you are treating: a) continue to apply both currents, b) turn off the HF and continue to apply the DC, c) turn off the DC and continue to apply the HF.

25. Large, fast-flowing vessels may be better coagulated if you: a) insert at a low angle and contact more of the vessel lumen, b) insert at a 90 degree angle into the center of the vessel, c) use higher DC current.

26. An extensive condition, with multiple telangiectases, is often best treated: a) with DC only, b) with HF only, c) by pulsing the HF on and off, d) by keeping the HF and DC on at all times.

27. When removing a spider telangiectasia, you should: a) first remove larger radiating capillaries, then treat the raised center, b) treat the raised center only, c) treat the radiating capillaries only, d) insert at a low angle to treat both the center and radiating capillaries.

28. If you see blood leaking from the treated area: a) try to coagulate the leaking blood, b) gently mop up the leaking blood, but do not treat again, c) apply pressure to stop the micro-hemorrhage.

29. When treating a large protruding "blood mole" on the trunk, you can expect: a) difficulty in producing coagulation, b) fast blood flow, c) the lesion is a vein, d) large crusts will form after the treatment.

30. To avoid having the treated vessel reopen: a) stretch the skin as you work, b) gently press the skin toward the center of the face to avoid accidental stretching, c) press on the area to minimize bleeding.

31. With all patients: a) schedule a 15 minute initial treatment, b) wait 2 days between the first and second treatment, c) do a short preliminary treatment and wait about 2 weeks to assess treatment success.

32. Generally, you should: a) treat an area no larger than 1.5 sq cm, and separate treated areas by at least 2 cm, b) concentrate your insertions, c) scatter your work over the entire face.

33. Crusting after telangiectasia treatment: a) seldom takes place, b) always takes place—but may not be perceptible, c) indicates poor technique.

34. Post treatment pigmentation is caused when treatment stimulates: a) macrophages, b) melanosomes, c) melanocytes, d) the basal layer.

35. Post treatment pigmentary problems are rare because: a) most patients with telangiectases are fair-skinned, b) the blend never causes this condition, c) cataphoresis eliminates the problem.

CORRECT ANSWERS

Multiple Choice: 1-b, 2-b, 3-c, 4-a, 5-c, 6-a, 7-b, 8-b, 9-d, 10-b, 11-c, 12-d, 13-b, 14-a, 15-c, 16-d, 17-b, 18-a, 19-d, 20-c, 21-c, 22-c, 23-b, 24-c, 25-a, 26-d, 27-a, 28-b, 29-d, 30-b, 31-c, 32-a, 33-b, 34-c, 35-a.

True & False: 1-T, 2-F, 3-T, 4-F, 5-T, 6-F, 7-T, 8-T, 9-T, 10-F, 11-F, 12-T, 13-T, 14-T, 15-F, 16-F, 17-T, 18-F, 19-T, 20-F, 21-T, 22-F, 23-F, 24-T, 25-T, 26-T, 27-T, 28-T, 29-F, 30-T, 31-F, 32-T, 33-F, 34-T, 35-F, 36-T, 37-F, 38-F, 39-T, 40-T, 41-T, 42-F, 43-T, 44-T, 45-F.

True & False Questions

1. The term telangiectasia originates from Greek and literally means "end vessel dilation."

2. Injury to the skin that causes bleeding and bruising often results in the formation of telangiectases.

3. A protruding "blood mole" may be found on the chest or back.

4. Telangiectasia is a disease.

5. Overexposure to sunlight can cause telangiectases in predisposed people.

6. Dental and chest X-rays are known to cause telangiectases.

7. Telangiectases occur more frequently in women.

8. Oral contraceptives that contain estrogen can elevate hormone levels and cause telangiectases.

9. High blood pressure is known to cause telangiectases in some patients.

10. Telangiectases do not help determine the presence of specific diseases.

11. If a patient has telangiectases, yet appears healthy, AIDS may be the cause.

12. Scleroderma typically produces telangiectasia.

13. Most patients with telangiectases have minor cases that are not disease-related and are of cosmetic importance only.

14. In certain people, capillaries of the nose and cheeks dilate from psychological stimulation.

15. A person with a bright red nose is obviously an alcoholic.

16. Telangiectasia is seldom a family trait.

17. Inherited telangiectasia occur in up to 15 percent of the completely normal healthy population.

18. Improper telangiectasia treatment by coagulation can lead to embolism in the heart or lungs.

19. After being coagulated, the vessel is replaced by scar tissue.

20. In some cases, proper removal of telangiectases can "starve" the skin and cause skin necrosis.

21. A venule is a small vein.

22. The blend works well in removing large varicose veins of the legs.

23. With the blend method, tiny leg venules are often eliminated in one treatment.

24. Skilled physicians can remove large veins in the legs using injectable sclerosant solutions.

25. For most patients, the key disadvantage of using laser is the high cost of treatment.

26. To correct the problem of the needle sticking, HF operators use the so-called "tapping technique."

27. Using the **basic procedure** blend technique, the HF coagulates a segment of the telangiectasia.

28 Using the **basic procedure**, the DC is kept on as the needle is withdrawn, so that the needle does not stick to the coagulated segment.

29. When treating a vessel, if you see bubbles forming in the capillary from the DC, the blood in the vessel is usually moving slowly.

30. If a vessel is fast-flowing, you may need higher HF to coagulate the blood.

31. The center of a spider telangiectasia is raised because the vein carries blood out into the capillaries.

32. Insertion into a spider telangiectasia is usually deeper than for simple linear telangiectases.

33. If blood is seen leaking from the treated telangiectasia and pooling under the epidermis, the treatment has been a failure.

34. A simple red dot may appear on the nose or lips, and is called a dot-telangiectasia.

35. If you see red blotchy areas, with no clearly defined vessels, you may not treat the area.

36. Minuscule telangiectases can sometimes be effectively eliminated using low DC only.

37. It is a good idea to have a physician administer local anesthetic for telangiectasia work.

38. The patient should take aspirin before treatment to subdue treatment pain.

39. As you treat the area, swelling takes place and hides untreated telangiectases.

40. Overheating the skin by too long a treatment can reopen the coagulated vessels.

41. Cataphoresis may be used after the treatment at about 0.5 to 1.0 milliamperes.

42. To ensure treatment success, the patient should periodically stretch the skin following treatment.

43. Post-inflammatory hyperpigmentation is common in dark-skinned patients following telangiectasia treatment with the blend.

44. You may recommend sunscreen lotion to reduce the recurrence of telangiectases in predisposed patients who live in states with sunny climates.

45. Experts believe that vitamin C is ineffective in strengthening blood vessel walls.

REGISTRATION

DON'T FORGET TO REGISTER TO RECEIVE UPDATES!

By registering with the Blend Educational Network, you will receive updates on information dealing with telangiectasia treatment—including seminars.

You will also receive notification of new publications dealing with related subjects.

We also plan to make available a business package for telangiectasia promotion, including: form letters, pricing lists, office brochures, patient after-care instructions, etc.

FILL OUT THE CARD BELOW AND MAIL TO THE FOLLOWING ADDRESS:

Blend Educational Network
1310 San Miguel Avenue
Santa Barbara, California 93109-2043 U.S.A.

REGISTER MY NAME WITH THE BLEND EDUCATIONAL NETWORK...

NAME: _____

STREET ADDRESS: _____

CITY AND STATE: _____

ZIP CODE: _____

TELEPHONE:

COUNTRY:

You may cut out the above card, photocopy, or just send in your name, address and telephone number. Thank You!